The Nature and Role of Algebra in the K-14 Curriculum

Proceedings of a National Symposium May 27 and 28, 1997

Sponsored by

National Council of Teachers of Mathematics and

Mathematical Sciences Education Board

Center for Science, Mathematics, and Engineering Education

National Research Council

National Academy Press
Washington, D.C. 1998

NATIONAL ACADEMY PRESS • 2101 Constitution Avenue, NW • Washington, DC 20418

NOTICE: The project that is the subject of this report was approved on February 21, 1997, by the Executive Committee of the Governing Board of the National Research Council (NRC), whose members are drawn from the councils of the National Academy of Sciences, the National Academy of Engineering, and the Institute of Medicine. The project was approved on April 22, 1996, by the Board of Directors of the National Council of Teachers of Mathematics (NCTM).

This report has been reviewed by a group other than the authors according to procedures approved by a Report Review Committee consisting of members of the National Academy of Sciences, the National Academy of Engineering, and the Institute of Medicine.

Additional copies of this report are available from the National Academy Press, 2101 Constitution Avenue, NW, Lock Box 285, Washington, DC 20055. (800) 624-6242 or (202) 334-3313 (in the Washington metropolitan area).

International Standard Book Number 0-309-06147-4

This report is available online at <http://www.nctm.org> and at <http://www.nap.edu>

Printed in the United States of America.

NATIONAL COUNCIL OF TEACHERS OF MATHEMATICS

The National Council of Teachers of Mathematics (NCTM), founded in 1920, is a nonprofit professional association dedicated to the improvement of mathematics education for all students in the United States and Canada. It offers vision, leadership, and avenues of communication for those interested in the teaching and learning of mathematics at the elementary-school, middle-school, high-school, college, and university levels. With more that 110,000 members, NCTM is the largest mathematics education organization in the world. Each year, the NCTM conducts a large national conference and seven to nine regional conferences, where teachers of mathematics and others interested in mathematics education can attend lectures, panel discussions, and workshops and can see exhibits of the latest mathematics education materials and innovations. Many NCTM members are also members of one or more of the 260-plus local and special-interest groups formally affiliated with NCTM that work in partnership with the Council to meet mutual goals. As a professional association, the NCTM derives its strength from the involvement of its members, who are drawn from the broad community of stakeholders interested in the field of mathematics and mathematics education.

NATIONAL ACADEMY OF SCIENCES • NATIONAL RESEARCH COUNCIL • CENTER FOR SCIENCE, MATHEMATICS, AND ENGINEERING EDUCATION • MATHEMATICAL SCIENCES EDUCATION BOARD

The National Academy of Sciences (NAS or the Academy) is a private, nonprofit, self-perpetuating society of distinguished scholars engaged in scientific and engineering research and dedicated to the furtherance of science and technology and to their use for the general welfare. Upon the authority of the charter granted to it by the Congress in 1863, the Academy has a mandate that requires it to advise the federal government on scientific and technical matters. The National Research Council (NRC or the Council) was organized by the National Academy of Sciences in 1916 to associate the broad community of science and technology with the Academy's purposes of furthering knowledge and advising the federal government. Functioning in accordance with general policies determined by the Academy, the Council has become the principal operating agency of both the National Academy of Sciences and the National Academy of Engineering in providing services to the government, the public, and the scientific and engineering communities. The Council is administered jointly by both Academies and the Institute of Medicine.

The Mathematical Sciences Education Board (MSEB) was established in 1985 by the National Research Council to maintain a national capability for assessing the status and quality of mathematics education. The MSEB is located within the Center for Science, Mathematics, and Engineering Education (CSMEE or the Center), which was established in 1995 to provide coordination of the NRC's education activities and reform efforts for all students at all levels, specifically those in kindergarten through twelfth grade and in undergraduate, school-to-work, and continuing education programs in the disciplines of science, mathematics, technology, and engineering. The Center reports directly to the Governing Board of the NRC.

Acknowledgments

The National Council of Teachers of Mathematics (NCTM) and the Mathematical Sciences Education Board (MSEB) both have as their mission the improvement of mathematics education. Because algebra is one of the cross-cutting content areas in the mathematics curriculum and a topic that is currently of much concern to the mathematics community at large, the two organizations joined together to organize a national symposium on algebra in May of 1997. This joint venture was unique in the relationship between the two organizations and represents a significant step in bridging the diverse communities represented by the two organizations. The symposium was organized by the Algebra Symposium Task Force of NCTM and a subgroup of MSEB members.

We gratefully acknowledge the National Science Foundation (NSF), whose financial support (Award #9614977) made the symposium possible, and Texas Instruments and Casio, who provided additional funds. Any opinions, findings, and conclusions or recommendations expressed in this material are those of the authors and do not necessarily reflect those of the funders.

We also would like to acknowledge the staff at NCTM and at MSEB for their efforts in putting the symposium together. In particular, Ramona Irvin from MSEB and Nancy Hawthorne from NCTM spent many hours drawing up invitation lists, contacting participants, arranging housing, and, in general, ensuring that the details were in place for a successful meeting. They were supported in their efforts by Catherine Bell and Colleen McGurkin from MSEB and Kathleen Chapman and Mary Ferris from NCTM. A special thank-you goes to Marilyn Hala from NCTM for shepherding the grant-writing process and for her help on-site during the symposium. Others who provided on-site support were Francis (Skip) Fennell and Bradford Findell from the MSEB staff and Virginia Williams, Joan Armistead, and Kathleen Chapman from the NCTM staff.

We are grateful to the speakers for their contributions and leadership that gave substance to the discussion and to Mark Saul and Bill Tate for providing thoughtful pre-conference readings. Finally, we would like to thank Francis (Skip) Fennell, James Gates, Kathleen (Kit) Johnston, and Beth Wallace from MSEB for their work in organizing and editing these proceedings for review and publication. A special thanks goes to Bradford Findell for his review of the mathematics in this work.

It should be noted that these proceedings have been reviewed by individuals chosen for their diverse perspectives and technical expertise, in accordance with procedures approved by the National Research Council's (NRC) Report Review Committee. The purpose of this independent review is to provide candid and critical comments that will assist the NRC in making the published report as sound as possible and to ensure that the report meets institutional standards for objectivity, evidence, and responsiveness to the study charge. The content of the review comments and draft manuscript remains confidential to protect the integrity of the deliberative process.

We wish to thank the following individuals for their participation in the review of this report:

Dr. Christian Hirsch, Western Michigan University
Dr. Roger Howe (NAS), Yale University
Dr. Henry O. Pollak, retired
Dr. Cathy L. Seeley, University of Texas
Ms. Bonnie Walker, Texas ASCD

While the individuals listed above have provided many constructive comments and suggestions, responsibility for the final content of this report rests solely with the authoring committee and the NRC.

GAIL BURRILL
President
National Council of Teachers of Mathematics

JOAN FERRINI-MUNDY
Director
Mathematical Sciences Education Board

NATIONAL COUNCIL OF TEACHERS OF MATHEMATICS
ALGEBRA SYMPOSIUM TASK FORCE

Beverly Williams, Chair
Pulaski County Special School District
Sherwood, Arkansas

Hyman Bass*
Columbia University
New York, New York

Laurie A. Boswell
Profile School
Bethlehem, New Hampshire

Sadie C. Bragg*
Borough of Manhattan Community College
The City University of New York
New York, New York

Gail F. Burrill
University of Wisconsin
Madison, Wisconsin

Leigh Childs
San Diego State University
San Diego, California

Shari Ann Wilson Coston*
Arkansas Education Renewal Consortium
Arkadelphia, Arkansas

Robert L. Devaney
Boston University
Boston, Massachusetts

Irvin E. Vance
Michigan State University
East Lansing, Michigan

Bert K. Waits
Ohio State University
Columbus, Ohio

---

*Mathematical Sciences Education Board members subgroup

MATHEMATICAL SCIENCES EDUCATION BOARD
(IN MAY OF 1997)

Preface

The National Academy of Sciences was pleased to host on May 27 and 28, 1997, a national symposium on "The Nature and Role of Algebra in the K-14 Curriculum" jointly sponsored by the National Council of Teachers of Mathematics (NCTM) and the National Research Council's Mathematical Sciences Education Board (MSEB).

One of the Academy's greatest strengths lies in its ability to act as a convener. As I observed the symposium on its last day, I heard and saw the breadth of the representation—across grade levels and across states. NCTM and MSEB clearly succeeded in their ongoing commitment to bring together thoughtful members of the mathematical sciences community to consider important questions in mathematics education. In this case, the questions involved the timely topic of algebra and how it should be treated in the K-14 grades.

This record of the symposium proceedings reflects the diversity of the symposium's speakers and participants. It is rich with the shared information and perspectives of elementary-, middle-, and high-school teachers, postsecondary and research mathematicians, teacher educators, mathematics education administrators, and others.

As you read the papers, presentations, and discourse of the symposium's two days here at the Academy, you will see that the subject of the nature and role of algebra in the K-14 curriculum is difficult and complicated. The questions that are being asked include, What do we mean by algebra and algebraic thinking? What do American students really need to know about and be able to do with algebra? How can we better prepare K-14 teachers to teach algebra? How can we better communicate to parents, the business community, and the general public about the kind of algebra that is relevant and why?

As the mathematics and mathematics education communities work with these questions, the Academy, the National Research Council, and the MSEB will continue to be active participants and partners. This is a high-stakes matter, and it will take all of our efforts to make sure that our nation and our nation's children are mathematically prepared for the 21st century.

Bruce Alberts
President
National Academy of Sciences

Contents

The Nature and Role of Algebra in the K-14 Curriculum

Introduction

With the 1989 release of *Everybody Counts* by the Mathematical Sciences Education Board (MSEB) of the National Research Council and the *Curriculum and Evaluation Standards for School Mathematics* by the National Council of Teachers of Mathematics (NCTM), the "standards movement" in K-12 education was launched. Since that time, the MSEB and the NCTM have remained committed to deepening the public debate, discourse, and understanding of the principles and implications of standards-based reform. One of the main tenets in the NCTM *Standards* is commitment to providing high-quality mathematical experiences to all students. Another feature of the *Standards* is emphasis on development of specific mathematical topics across the grades. In particular, the *Standards* emphasize the importance of algebraic thinking as an essential strand in the elementary school curriculum.

Issues related to school algebra are pivotal in many ways. Traditionally, algebra in high school or earlier has been considered a gatekeeper, critical to participation in postsecondary education, especially for minority students.[1] Yet, as traditionally taught, first-year algebra courses have been characterized as an unmitigated disaster for most students.[2]

There have been many shifts in the algebra curriculum in schools within recent years. Some of these have been successful first steps in increasing enrollment in algebra and in broadening the scope of the algebra curriculum. Others have compounded existing problems. Algebra is not yet conceived of as a K-14 subject. Issues of opportunity and equity persist. Because there is no one answer to the dilemma of how to deal with algebra, making progress requires sustained dialogue, experimentation, reflection, and communication of ideas and practices at both the local and national levels. As an initial step in moving from national-level dialogue and speculations to concerted local and state level work on the role of algebra in the curriculum, the MSEB and the NCTM co-sponsored a national symposium, "The Nature and Role of Algebra in the K-14 Curriculum," on May 27 and 28, 1997, at the National Academy of Sciences in Washington, D.C.

The goals of the symposium were

1. to promote an informed dialogue on issues concerning the K-14 algebra curriculum, such as
   - necessary changes in the current curriculum
   - identification of research needed to help build a continuous and effective algebra curriculum

---

[1]Hawkins, B.D. (1993.) "Math: The Great Equalizer— Equity 2000 and QUASAR, Improving Minority Standing in Gatekeeper Courses." *Black Issues in Higher Education*, 10(6), 38-41.

[2] Steen, L. (1992.) "Does Everybody Need to Study Algebra?" *Basic Education*, 37(4), 9-13.

- the role of technology in the teaching and learning of algebra and how technology can enhance the development of algebraic reasoning and conceptual understanding
- ways to change the public perception of algebra from a year-long course to a continuous K-14 mathematical experience
- issues surrounding all students and the learning of algebra and
- determination of the algebraic ideas that are important at different grade levels and the incorporation of these ideas into the curriculum

2. to provide examples of students' algebraic thinking, to synthesize research, and to consider how these factors can impact the reality of algebra in the school curriculum
3. to provide a forum for those involved in algebra-related curriculum projects at the elementary-school, middle-school, secondary-school, and postsecondary school levels to share their visions of curricula, teaching, and assessment
4. to prepare participants to replicate similar symposia on a smaller scale in their local regions.

Over 300 invited participants representing all the states attended the symposium, including K-14 mathematics teachers, mathematics supervisors, postsecondary educators, community college mathematicians, collegiate mathematicians, and representatives of the mathematics associations.

This report describes the symposium, including presentations and discussions. To summarize briefly, on the first day, Hyman Bass, Department of Mathematics, Columbia University, opened the meeting with a mathematician's perspective on algebra. He argued for an early emphasis on use of the real number line and the geometric representation of arithmetic operations. He emphasized the significance of the Euclidean algorithm, both as a source of place value representation of numbers and as an illustration of the similarities and differences between integer and polynomial arithmetic. John Dossey, mathematics educator, Illinois State University, then discussed algebra in the past and in the present, including changes outlined in a discussion document that is printed here as Appendix E. Dr. Dossey then presented several models to consider in striving for student understanding of algebra and discussed several goals for school algebra programs.

During the balance of the first day of the symposium, participants focused on key elements of algebra instruction by rotating through sessions entitled "Algebra: A K-14 Curriculum Endeavor"; "Technology and Algebraic Reasoning"; and "Algebra: What All Students Can Learn." In these first-day sessions, several speakers presented their view of the issues, and their papers are contained in this record of the proceedings. Participants then worked in groups to discuss questions with respect to the issues raised by the topics. The following questions were among those discussed:

- In order to develop coherence and depth in the algebra curriculum across K-14
  - —what transitions do students need to make and over what period of time?
  - —how long must students be engaged with an important mathematical idea so that understanding is achieved?
- In order to develop criteria or lenses to select curriculum, appropriate pedagogy, and assessment
  - —how should algebraic reasoning be defined?
  - —how should state, national, or college placement exams be taken into account?
- In order to develop long-term support for teachers and *Standards*-based algebra
  - —how much help do teachers need?
  - —how do we develop public support for teachers and *Standards*-based algebra?

The second day of the symposium focused on "Algebra in the Curriculum." The following topics were featured—functions and relations; modeling; structure; and language and representation. Individuals affiliated with National Science Foundation (NSF)-funded curriculum projects in elementary-school, middle-school, high-school, and postsecondary mathematics gave presentations on these topics. Their papers are included in this record of the proceedings. The afternoon session also brought together participants from common geographic regions to plan local and regional algebra initiatives, which are now under way. Closing remarks were provided by Beverly Williams, Chair of the Task Force that planned the symposium.

Several background readings were provided for participants, and this record of the symposium publishes them in full. Two invited papers were commissioned and are also included, in Appendix D: "Stating the Obvious: Mathematics Course Taking Matters," which raises equity issues around course taking and algebra; and "Algebra, Technology, and a Remark of I.M. Gelfand," which focuses on technology and algebra. A discussion document adapted from work done by the NCTM Algebra Working Group also was sent to participants prior to the symposium and appears here in Appendix E by permission of the NCTM. It is entitled "A Framework for Constructing a Vision of Algebra: A Discussion Document."

*Opening Remarks**

SUZANNE WOOLSEY: "Ladies and gentlemen, I can tell the mathematical community is a very comfortable and friendly community. My name is Sue Woolsey. I am the Chief Operating Officer of the National Academy of Sciences and the National Research Council.

On behalf of Bruce Alberts, the President of the National Academy of Sciences, I am here to welcome you formally and officially to this very important algebraic symposium.

We are delighted to be able to host this event, and we are especially excited to be partners with the National Council of Teachers of Mathematics, well known to be both the generalissimos and the shock troops in the reform of mathematics education.

The Mathematical Sciences Education Board at the NRC has been a very important part of our operations in terms of convincing the scientific and mathematical community of the importance of focusing a great deal of attention on elementary, secondary, and undergraduate mathematics and science. We are absolutely delighted that things have progressed so well. We realize there is an immense amount of work still to be done, so it is exciting to be able to participate with you in this symposium.

I would like to take a second to tell you a little bit about this building and to encourage you during the time you are here to spend a few minutes looking at the various symbolic aspects of it.

Outside of the Auditorium is a written history that you can take home with you of the construction of the Academy building. It contains a plot of the ceiling of the Great Hall, which was painted by a woman named Hildreth Meiere. It incorporates symbology of science and mathematics from Egypt, Greece, and Rome and ancient academies of sciences in the western hemisphere.

I think you will find the Great Hall interesting. It gives you a feeling for many aspects of society back when it was built, in 1924.

Let me now say that this symposium officially has begun. I would like to turn this event over to the president of the National Council of Teachers of Mathematics, Gail Burrill."

GAIL BURRILL: "Thank you, Sue. It is an exciting time for all of us to be here. It is an exciting time in mathematics education. I am sure that all of you are aware of President Clinton's initiative to have a voluntary national test for eighth graders in the year 1999, and while we all knew that mathematics education—and algebra in particular—were topics of great interest, this initiative of the President's has pushed the issues to the fore and made

---

*The following remarks by Suzanne Woolsey and Gail Burrill were edited on the basis of verbatim transcriptions.

them even more important. Part of the concern now is how we deal with algebra on this test and what this does to the curriculum. These kinds of questions make what we are about here even more important.

Algebra is actually just the tip of the iceberg—just part of the broader picture of mathematics education that we are thinking about—because we are going to have to face some of the same issues in terms of geometry and statistics and measurement. First, though, we have many questions before us about algebra—algebra for all, algebra in the eighth grade, how we factor in technology, the Third International Mathematics and Science Study (TIMSS) and what it says about how we teach algebra, algebra in the 12th grade, algebra in the first and second grades, algebra at the university level, algebra in community colleges.

One thing that I hope comes out of this is the beginning of a coherent vision of what algebra ought to be and look like from grades K through 14. It is exciting because here, at this meeting, we have people from all over the United States and Canada to consider this important topic of algebra.

We owe the National Science Foundation a great deal of thanks for sponsoring this symposium by giving a grant to the National Council of Teachers of Mathematics.

We also owe a great deal of thanks to Texas Instruments and to Casio, who have donated money to NCTM to support some of the costs associated with putting on this meeting.

On behalf of NCTM, I am very glad to welcome you and to say that I hope we can make this an extremely productive and useful two days. I hope this symposium will advance the national dialogue on how we make a difference in what we do with algebra.

I would like to introduce our first speaker today, Hy Bass. Hy is a member of the Mathematics Department at Columbia University and also the Chair of the Mathematical Sciences Education Board, which is housed in the National Academy of Sciences. Hy is going to help us get a glimpse of what algebra looks like from the mathematician's perspective."

Keynote Addresses

Algebra with Integrity and Reality

Hyman Bass
Columbia University
New York, New York

I come to you as a university mathematician who does research in algebra and who is interested in improving mathematics education in the schools. While I have lived professionally with the subject we call algebra, I am not an expert on what algebra means and looks like in the school curricula. Indeed, there does not seem to be much consensus about this. Much of the discussion here seems to envisage broad areas of mathematics, such as functions or modeling, and to identify within them certain algebraic ideas and techniques. I choose instead to focus on some cognitive issues in algebra and on our sense of the number systems that are at the root of algebra. This focus is germane to teaching in the elementary grades on up.

In preparing this talk, I have benefited greatly from communications with Deborah Ball, Joan Ferrini-Mundy, and Mark Saul, to whom I extend my thanks and appreciation.

WHAT IS ALGEBRA?

To begin, I shall adopt a working definition of algebra that may strike you as traditional.

School algebra—and **the root of all algebra**—is about the following:

- The basic **number systems**—the **integers** and the **real numbers** and those derived from them, such as the rational and complex numbers.
- The **arithmetic operations** (+, –, x, ÷) on these number systems.
- The **linear ordering** and resulting **geometric structure** defined on the real line. By this I mean the notions of size (whether one number is larger or smaller than another) and of distance between numbers.
- The study of the **algebraic equations** that arise naturally in these systems.

In some sense, *the subject grows from its efforts to solve equations.* The starting point, rooted in our experience with counting, is the **natural numbers**: 0, 1, 2, 3, 4, 5,... .[1] Then, trying to solve equations such as $8 + x = 5$, involving only natural numbers, leads us to introduce negative numbers and, so, to arrive at the **integers**: ..., –2, –1, 0, 1, 2, 3, 4, 5,... . Then, in order to solve equations such as $2x = 3$, with integer coefficients, we are led to introduce fractions, i.e., the **rational numbers**: numbers of the form p/q with p and q ($\neq 0$) integers.

[1] I am following Bourbaki in including "0" in the natural numbers, and, in fact, there are some mathematical reasons for doing this. There is divergent usage, however.

Many natural equations, such as $x^2 = 2$, cannot be rationally solved. We could continue enlarging our number system accordingly. However, even if we adjoin the solutions of *all* polynomial equations with rational coefficients, we still won't capture all of the real numbers.

The passage from the rational numbers to the real numbers is not via solutions to equations but, rather, by a very different process of "geometric completion," a sort of "filling in the holes" left by the rational numbers in the "real line." This process is not discussed explicitly in the schools and not even in most college mathematics courses. One of the main issues I want to address here is, *What sense of the real numbers is appropriate and useful for school mathematics?*

I want to advocate the following "definition," about which I'll say more later: **real numbers** are the points on the real number line, as shown here:

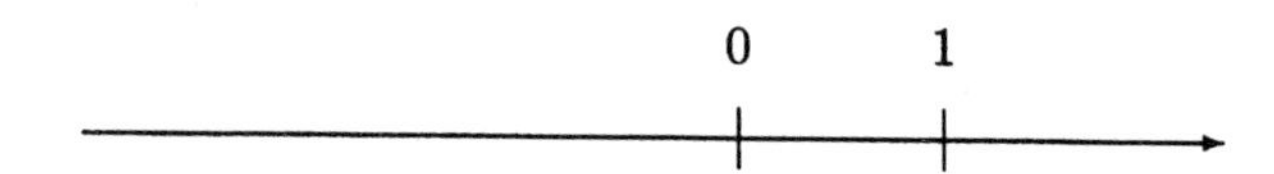

With real numbers, we select an origin, 0, and a unit, 1, to the right of the origin, and then each point is measured by its (oriented) distance from the origin, counted negatively if it is to the left of 0. In particular, by counting off multiples of the unit distance, we are able to identify integers with (evenly spaced) points on this real number line.

Squares of real numbers are positive, so equations like $x^2 + 1 = 0$, have no solutions in the real numbers. To solve them, we are led to introduce the **complex numbers**. These are numbers of the form $z = x + iy$, where x and y are real numbers, and i denotes a square root of -1. We often identify z with the point (x,y) in the coordinate plane and so identify the set of complex numbers with the plane, the real numbers then forming the x-axis.

The so-called "**Fundamental Theorem of Algebra**" says that we can now stop here, in the following sense: *Any algebraic (i.e. polynomial) equation with complex coefficients can be solved with complex numbers.* [2]

This is not, of course, the end of the story of equations. The complex numbers simply give us a stable environment in which solutions live. We can still ask how deeply into the complex numbers we have to go to find solutions for particular types of equations and how we can describe them. For example, we know the familiar quadratic formula for the solutions of quadratic polynomial equations. Is there a similar formula, involving only rational operations and extraction of roots (square roots, cube roots, and so on) that describes solutions of higher degree polynomial equations? After much searching, the answer turns out to be "no," starting with degree five. The methods invented to prove this gave rise to a major branch of algebra called "group theory." The most important recent successful analysis of algebraic equations came with Andrew Wiles' celebrated proof of Fermat's Last Theorem.

The complex plane is a very attractive mathematical object. The arithmetic of the complex numbers gives a natural and integrated setting in which to introduce some linear algebra and in which to study the Euclidean transformations of plane geometry, an attractive possibility for the high-school curriculum. This topic deserves a paper of its own. But I want to focus here on some more fundamental, cognitive issues.

OUR SENSE OF THE NUMBER SYSTEMS

Suppose that I ask you, "What is a number?" The answer might be an instance of it, a visual or symbolic representation, a mental image .... What I want to emphasize also are major cognitive distinctions between thinking of individual numbers and thinking of "number systems." For natural numbers and integers, the answer to my question is pretty clear and natural, and it is well reflected in early school mathematics. Counting activities and discovery of number patterns with integers prevail. Numbers are represented as counts of collections of objects (apples, rods, tiles, for example). These settings provide models of addition and multiplication, and the integers are listed in order on an integer number line. The mental sense of the number 5 might progress from images of 5 apples or 5 children or 5 circles or 5 marks to a place labeled "5" on the integer number line. For very large numbers, the

[2]Despite the use of the words "theorem" and "algebra," this is not really a theorem in algebra, since its proof has to invoke non-algebraic ideas from either calculus or topology.

concrete images are no longer so intuitive, but the compressive power of place-value notation and the indefinite geometric extension of the integer number line can still bring them, at least implicitly, into one's mental universe.

What is the corresponding story for real numbers, and how are the two developments related? I asked myself, "*What, in the minds of school teachers and students, is a real number*?" This is hard to know, but one can at least look at textbooks and curricula. One of the explicit answers given is that "real numbers are the rationals and the irrationals." For this answer to be sensible, there has to be an implied understanding of some "number universe" in which the "irrationals are the complementary set to the rationals." According to the above definition, the real numbers would constitute this implied number universe. But has this universe been explicitly described? Until it is, the question of what the real numbers are has not really been answered.

If one looks at curricula for an operational definition of real numbers, they seem to appear ultimately, and only tacitly, as numbers represented by decimals—as a kind of by-product of the division algorithm for integers. This definition has some mathematical body, but it is technical, not intrinsic, and it poses some computational problems, as we shall see. The answer that I want to propose as appropriate for school mathematics is that *a real number is a point on the real number line*.

You may well ask, "What do you mean by the real number line?" My answer is that I view this as one of the primordial objects of mathematical experience, on which we can build our early mathematical learning. Just as whole numbers are rooted in our primitive experience of counting and enumeration of collections of discrete objects, so too—slightly later—do we have qualitative experience with (continuous) geometric measurement—of height, distance, speed, for example—and in such activities as cutting paper models, sewing, simple carpentry, and so on. This gives us a natural intuitive access to and geometric sense of the real line as a scale of measurement, even before we have a developed notational system for designating individual real numbers. Our geometric intuition is a sound and coherent foundation on which to build a geometric discussion of the basic arithmetic operations with real numbers, as I shall try to illustrate. And I suggest, also, that there are significant advantages to giving this geometric model of the real numbers an earlier and more prominent appearance in our school curricula.

First let us look at how the American curriculum develops the number systems. As far as I can tell, early mathematics instruction begins with counting numbers. Addition is modeled by aggregating collections of objects: $5 + 7$ is represented by a basket of 5 apples, combined with a basket of 7 apples. Multiplication is modeled by iterated addition: 5×7 is represented by 5 baskets, each with 7 apples. In the course of this, one learns the base-10 place-value notation for numbers and develops the conventional algorithms first for addition and then multiplication of such numbers first by memorizing single digit multiplication and then by learning an algorithm that reduces general products to a sequence of steps, each being a single digit multiplication or an addition. Fractions are then introduced and also division and decimal notation. The rules for arithmetic are extended to fractions and finite decimals, in each case as though these were new situations. When students are taught to divide integers, they encounter infinite decimals. In this way, finally, the real numbers, as arbitrary decimals, appear unannounced from the wings, almost as a kind of by-product of the algorithm for division in place-value notation.

If this is how real numbers emerge in the American curriculum, it presents some mathematical difficulties. For example, the conventional algorithms for both addition and multiplication begin with operations on the rightmost digits in decimal form. What can this mean for infinite decimals? This can be addressed mathematically, of course, but it doesn't seem to be. In fact, the very existence and properties of multiplication of real numbers appear to be unspoken assumptions, given as a part of some canon of unexplained origin or perhaps simply provided as a calculator command.

To illustrate the mathematical problem, contrast the understanding of arithmetic with fractions with that of real numbers. For example, multiplying by $1/2$ is understood to represent cutting, or dividing, something in half. But how is one to interpret multiplying by $\sqrt{2}$ or by π? Yet students will comfortably discuss expressions involving xy, where x and y are tacitly allowed to vary over all real numbers. It seems to me that the most satisfactory way to define multiplication for *all* real numbers is in terms of areas of rectangles, using the geometric model of the real numbers.

The present curricular approach, almost exclusively tied to base-10 place-value and decimal notation for numbers, does not support a conceptually well-grounded sense of the real number system. Missing is an early and systematic emphasis on the geometric representation of the (continuous) real line and, correspondingly, on the geometric representation of the operations of arithmetic.

The classical Greek geometers did not have the benefit of our place-value notation but did have a deep sense of the real number system and focused a great deal of attention on number theoretic questions, e.g. the existence and irrationality of certain square roots and the nature of the number π. For the ancient Greeks, real numbers were measures of geometric quantities (length, area, and so on). It is in that same spirit that I propose that we found our sense of the real numbers on the geometric real line model (see below), in which we first name the integer points and eventually name all of the points with the aid of decimal notation. Of course, there is the serious question about whether what I am advocating can be made accessible to young students.

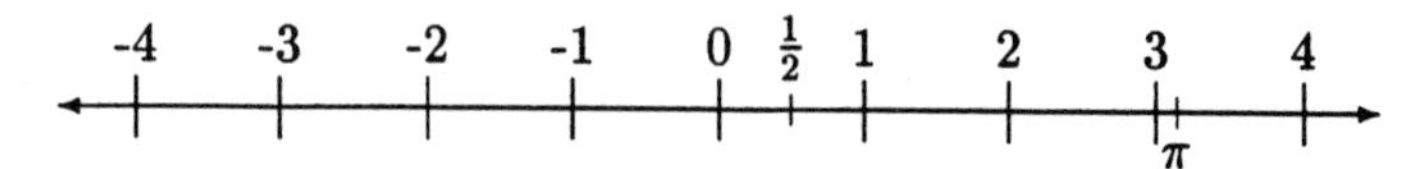

What's the point? What does one gain by this? First of all, one has an intuitively accessible and authentic model of the real numbers that can be apprehended at a very early age. Second, the model's geometric representation is completely intrinsic and not dependent on the structure of place-value or any other arbitrary notation. Third, the basic operations of +, –, and × have very natural geometric interpretations, making some properties transparent when they are not at all in the combinatorial or counting model. For example, addition can be done by juxtaposing intervals; i.e., $a + b$ is obtained by placing the left end of an interval of length b at the right end of one of length a, as shown by the first figure below. Subtraction, $a - b$, can be modeled by making the right ends of the intervals coincide and then locating the left end of the b-interval, as shown by the second figure below.

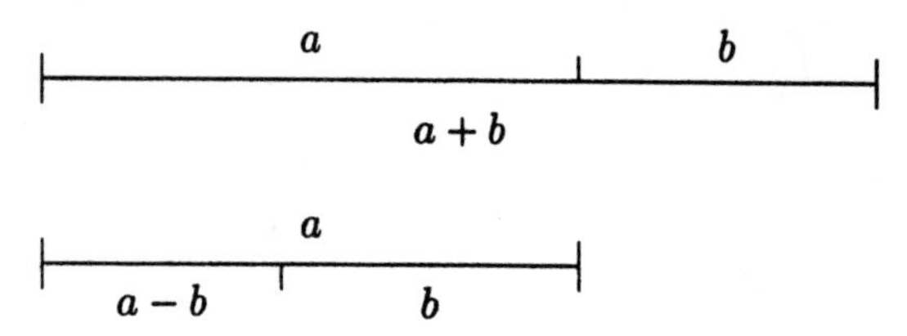

Multiplication, in contrast, *is two dimensional*; i.e., $a \times b$ is the *area* of a rectangle with sides a and b. Commutativity is made transparent, a fact not obvious in the combinatorial model: compare 5 baskets of 7 apples with 7 baskets of 5 apples. The rectangle model of multiplication also makes the distributive law transparent. Notice further that these geometric constructions directly, and intrinsically, define the operations for *all* real numbers, whether or not they have been named (e.g. by place-value notation).[3]

$a \times b = b \times a$ $\qquad$ $a \times b + a \times c = a \times (b + c)$

The process of adding or multiplying a fixed number with a variable number can be geometrically interpreted as a *transformation of the real line*. Adding 2 is a translation— distance 2 to the right. Adding –5 is a translation— distance 5 to the left. Multiplying by 3 is a tripling of scale. Multiplying by 1/4 is a reduction of scale to one-fourth size. Multiplication by –2 is a doubling of scale, combined with a (mirror) reflection through the origin. It is this reflection interpretation of multiplication by negative numbers which demystifies the infamous formula, $(-a)(-b) = ab$, because double reflection brings you back to where you started.

Such illuminating discussion of arithmetic operations as geometric transformations of the line seems to be absent from the current curriculum until one investigates what happens to the graph of a function, $y = f(x)$, when you change x to ax, or $x + a$, or change y to ay, or $y + a$. But there is hardly need to wait for the discussion of function graphs to introduce this transformational point of view.

[3]There is a serious underlying problem here: explaining how one can convert a measure of area into a real number (i.e., a linear measure). This may require reversion to the repeated addition model but with grids of finer and finer units and then a passage to a limit. In the early grades, one could give only the most heuristic account of this.

If one reaches the complex numbers, then the same transformational interpretation of addition and multiplication (plus complex conjugation) leads to all of the fundamental Euclidean (rigid) motions of plane geometry, on which the notion of geometric congruence and similarity of figures is based.

This discussion has suggested how one could approach some of the early learning of arithmetic using the geometric model of the real numbers. Next we will look at some of the features of this model at a more advanced level of study.

EUCLIDEAN DIVISION

First, I am not arguing for a de-emphasis of place-value notation. After all, place-value is the basis for our most effective algorithms for arithmetic. Thus, if the above geometric point of view is to receive emphasis, it must be reconciled with place-value notation. This reconciliation is itself a very instructive undertaking and can be based on some of the fundamental properties of real numbers, which themselves perhaps deserve increased attention.

Place-value flows from a basic result, sometimes called "**Euclidean Division,**" which I believe deserves some emphasis, since, as I shall illustrate below, it has many other significant applications as well. The idea is as follows:

Given a number $b > 0$, we mark off the real line with "fence posts" at distance b apart, with one post at 0. These posts are then located at the set of integer multiples of b and look on the real line just like the integers but for a change of scale.

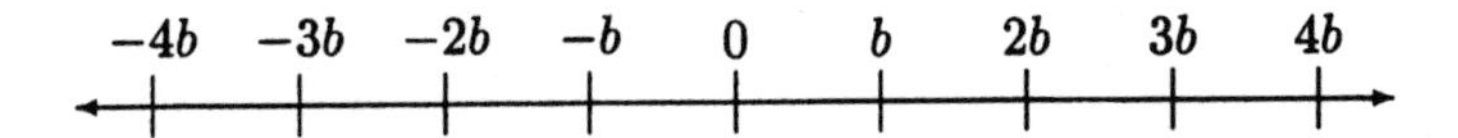

Now any given number will lie between two consecutive posts. If qb (with q being an integer) is the nearest post to the left of a, then we have this picture:

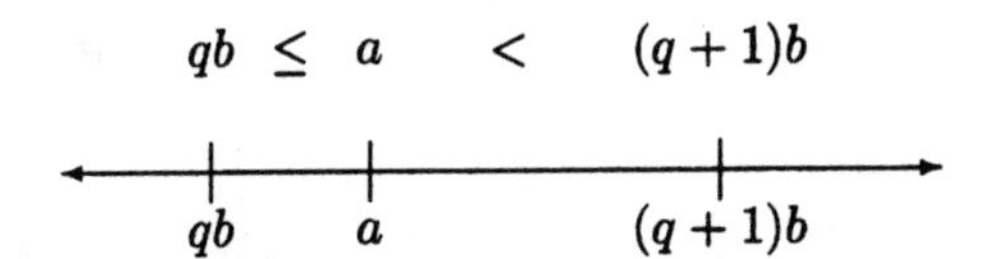

Subtracting qb as represented on the previous page, we have

$$0 \leq a - qb < b$$

Now, when we put $r = a - qb$, we have proved (heuristically) **Euclidean Division**, which is as follows:

- Fix a real number $b > 0$. Then for any real number a, there exist a unique integer q ("quotient") and real number r ("remainder"), $0 \leq r < b$, such that $a = qb + r$.

Note that if a and b are integers, then so also is $r = a - qb$.

OBSERVATIONS

Let me make some observations to explain the significance of this result.

1. This "proof" was purely geometric (and intuitive) and not dependent on any special notation for numbers, other than their geometric definition on the real line.

2. The result is related to ordinary division as follows: Dividing by b transforms the equation $a = qb + r$ to $a/b = q + r/b$, with q an integer and $0 \leq r/b < 1$. Here, q—called the "integer part" of a/b—is what appears to the left of the decimal point when we divide a by b, while r/b—called the "fractional part" of a/b—is, when written in decimal form, what appears to the right of the decimal point when dividing a by b.

For example, with $b = 123$ and $a = 1997$, we have $a/b = 1997/123 = 16.235772357723577...$, so $q = 16$ and r/b (= 29/123) = 0.235772357723577... .

3. Consider what happens for $b = 10$. For example, taking $a = 5{,}297$, we have $5{,}297 = 529 \times 10 + 7$, so $q = 529$ and $r = 7$, the units digit of a. Repeat this with $q = 529$ to get $529 = 52 \times 10 + 9$.

Substituting this in the first equation, we get $5{,}297 = (52 \times 10 + 9)10 + 7 = 52 \times 10^2 + 9 \times 10 + 7$.

Finally, writing $52 = 5 \times 10 + 2$, and substituting in the last equation gives $5{,}297 = (5 \times 10 + 2)\,10^2 + 9 \times 10 + 7 = 5 \times 10^3 + 2 \times 10^2 + 9 \times 10 + 7$.

Thus, repeated application of **Euclidean Division** to the successive quotients yields the *expansion of a as a sum of powers of 10, with base-10 digits (the numbers 0, 1, 2, 3,... , 9) as coefficients. This is exactly equivalent to the base-10 place-value representation of a.* The significance here is that we obtained it intrinsically from repeated application of Euclidean Division. Moreover, by exactly the same process, for *any* integer $b > 1$, we can so obtain the base-b place-value representation of any integer $a > 0$.

Further, if $a > 0$ is not an integer, then we can apply the above to the integer part of a, and then apply a related but modified procedure to the fractional part of a to derive the place-value representation of a to the right of the "b-ecimal" point.

While this is all a bit technical, I simply wish to emphasize that *Euclidean Division contains the genesis of the place-value representation of all numbers, to an arbitrary integer base* $b > 1$. This is one reason for its mathematical importance.

4. Given integers a and b, suppose that we want to calculate their *greatest common divisor,* gcd(a,b). A natural way to try to do this is first to calculate the prime factorization of a and b and then, for each prime p, to take the highest power of p dividing both of them, and, finally, to take the product of these prime powers. This turns out to be a computationally burdensome method. The task of factoring very large integers is extremely laborious, even with powerful computers. Indeed, the difficulty of this problem is the source of security in some of the commonly used cryptosystems.

On the other hand, one can quite simply calculate gcd(a,b) without knowing the prime factorizations, as follows. Say that $b < a$. Then we can (Euclidean) divide a by b to get a remainder $r < b$. If $r = 0$, then b divides a, and b = gcd(a,b). If $r > 0$, we then divide b by r and get a new remainder, $r' < r$. If $r' = 0$, then r = gcd(a,b). If $r' > 0$, we then can continue in the same fashion. This process must terminate in a finite number of steps, as the successive remainders keep getting smaller.

More formally, we may define the (decreasing) sequence of integers—$a_1, a_2, a_3, a_4,...$ by $a_1 = a$, $a_2 = b$, and, for $n > 1$, a_{n+1} is the remainder obtained when dividing a_{n-1} by a_n. This decreasing sequence of non-negative integers strictly decreases until it reaches 0 in a finite number of steps, and the last non-zero term is gcd(a,b).

Thus, ***Euclidean Division*** *gives us an algorithm for calculating the greatest common divisor,* gcd(a,b), *for any two integers a and b.* A further important consequence of this algorithm is an expression of the form, gcd(a,b) = $ra + sb$, *for suitable integers r and s.*

It is this latter result that underlies the proof of unique prime factorization for integers, i.e., the so-called **"Fundamental Theorem of Arithmetic."**

5. One of the most interesting and far-reaching analogies in mathematics is the similarity of the arithmetic of the integers and that of the "ring" of polynomials in one variable x, say with real or complex coefficients. The starting point of this analogy is Euclidean division of polynomials, where now the "size" of a polynomial is measured by its degree. Again, we have the following:

Given polynomials b and a, with $b \neq 0$, there exist unique polynomials q and r, so that degree(r) < degree(b), and $a = qb + r$.

The idea is to keep subtracting suitable multiples of b from a to keep lowering the degree, until one obtains a remainder of degree smaller than that of b. Formally this argument is carried out by mathematical induction on degree(a).

Now, just as for the integers, this leads to an algorithm for calculating the greatest common divisor of two polynomials and, as above, to a proof of unique factorization for polynomials.

6. There is a further interesting, formal similarity between place-value notation for integers and for polynomials. As noted above, base-10 place-value representation corresponds to expression of a number as a sum of powers of 10, with digit coefficients. For example, $5{,}297 = 5 \times 10^3 + 2 \times 10^2 + 9 \times 10 + 7$.

If we substitute "x" for "10" here, then this becomes a polynomial. So we can more or less think of place-value notation as expressing numbers as "polynomials in 10" instead of "polynomials in a variable x." But the analogy is not so simple, and it breaks down when one analyzes the rules for arithmetic. The way that this breaks down in fact illuminates some of the complications of our algorithms for arithmetic.

Consider first two polynomials, for example, $a = a_0 + a_1x + a_2x^2$, and $b = b_0 + b_1x$.

Then we have, from commutativity of addition and distributivity of multiplication, $a + b = (a_0 + b_0) + (a_1 + b_1)x + a_2x^2$, and $a \times b = a_0b_0 + (a_1b_0 + a_0b_1)x + (a_2b_0 + a_1b_1)x^2 + a_2b_1x^3$.

For polynomials, this is the end of the story. The reason is that the coefficients of polynomials, the scalars (integer, rational, real, or complex), are closed under addition and multiplication.

Now, in contrast, imagine in the equations above that we replace x by 10 and assume that the coefficients a_0, a_1, a_2, b_0, b_1 are base-10 digits. Then exactly the same formulas for addition and multiplication apply, since they are the result simply of commutativity and distributivity of addition and multiplication. The complication arises from the fact that new coefficients, such as $a_1 + b_1$ or $a_2b_0 + a_1b_1$, need no longer be base-10 digits; they may be larger than 10. Hence, to express the answers in standard place-value form, we have to remove some 10's, and "carry" a 1 to the adjacent coefficient on the next higher power of 10. This is the source of the "carries" in our standard algorithms for addition and multiplication.

SUMMARY

I have argued that it is both natural and advantageous to give an early emphasis to the geometric real line model of the real numbers, in which basic arithmetic operations are interpreted geometrically and developed alongside the more algorithmic development rooted in base-10 place-value representation.

I also have used **Euclidean Division** to illustrate several important algebraic phenomena. First, I have shown how it can be used to derive place-value representation from the geometric model of the real line. Second, you can see that, when applied to integers, **Euclidean Division** furnishes an effective algorithm for calculating greatest common divisors and the basis for proving the **"Fundamental Theorem of Arithmetic"**—unique prime factorization. Then, an analogue of Euclidean division for polynomials can be used to lead to analogous results, thus showing a "mathematically similar structure," so to speak, for the arithmetic of integers and the arithmetic of polynomials. (I want to note that searching for this kind of similarity of mathematical structure in different looking settings is characteristic of research in algebra and, indeed, in all mathematics in general.) Finally, I have offered some remarks on the similarities and differences between integer and polynomial addition and multiplication.

Making Algebra Dynamic and Motivating: A National Challenge

John A. Dossey
Illinois State University
Normal, Illinois

I have been given the task of discussing what high-school algebra is and what it should be. To lessen the suspense, I will tell you that I am not going to give you the answer as to what it should be. However, this is a question that deserves to be addressed in a place as august as the National Academy of Sciences. I cannot think of a more important time in the history of our country than now to focus efforts on the teaching and learning of algebra in our schools. As Hy Bass has shown, algebra has a long and beautiful history, but the translation of that into a school subject has lost some dynamism. For many students and teachers, algebra has become a mere passing on of information from generation to generation.

To prepare my remarks, I went back and looked at school books for algebra from more than a century ago. I found that they are not very different from my high-school algebra book and, in fact, not very different from some books that have 1996 copyrights. Charles Davies' algebra text (1846), which is a translation of the French version by M. Bourdon, begins with some definitions that focus on number systems. Next, it looks at polynomial expressions and solutions of first-degree equations. The focus then shifts to exponents and radicals, at least to the point you can talk about quadratic equations and what square roots might be. Davies then does a little factoring and solutions of quadratic equations and systems of linear equations. Does all that sound familiar to you?

As my text analysis moved toward the present, I noticed some differences. In some cases, there has been a permutation of topics. In some cases, the solution of first-degree equations appears before the discussion of polynomial expressions. These differences serve to highlight the shift in the treatment of the subject from meaning to manipulation. Overall, the text analysis showed the very static nature of the curriculum in secondary school algebra.

Discussing the role of algebra in the broader realm of the K-14 curriculum is an even more daunting task. After all, we are products of the system of algebra teaching that I have just outlined. When we talk about making changes in algebra teaching and learning, we are talking about changing the system of concepts, skills, and knowledge that provided part of the basis for our being here today. We are the survivors of that system and the sifting process it helped define. The very skills and knowledge that gave us our identifications as mathematicians and mathematics educators—those knowledgeable about mathematics and its teaching—were a result of our performance as we moved through "algebra."

Now, one might say, "We have the *Curriculum and Evaluation Standards for School Mathematics* now, and, so, things are changing." But what do we see when we look at the National Assessment of Educational Progress (NAEP) results and the Third International Mathematics and Science Study (TIMSS) reports from teachers about what they are doing in the algebra classroom? The data show a little bit of movement toward doing projects. They show a little bit of movement toward using technology. Involved mathematics educators immediately react to this stating that they know many teachers who really are changing. However, if you triangulate from the NAEP assessments, the TIMSS findings, and from dissertation research on teacher beliefs, you see a movement toward change that can be measured in terms of geologic time scales.

There has been change as a result of the *Standards*. It is of an undeniable and growing magnitude. But, it is not clear that it has taken a uniformly accepted direction. This growing vector of change may yet lead the way to a new algebra for school mathematics. We see the beginnings of change in algebraic related pedagogy. It is not clear that the underlying content to which this pedagogy is applied has really received due attention yet, however. Referring to M. Fullan's book (1993) on educational change, we probably are on course. We are now nearly 10 years past the release of the *Standards* and that is an appropriate time frame for the initiation of ideas. We now need to shift to the beginning of the implementation phase. The question is, what do we implement? We must take into consideration that beyond the stage of implementation is institutionalization. That is what we are suffering from right now. We are suffering from the institutionalization of a certain type of school algebra that is at least 200 years old.

What changes do we see in new programs and new books? In programs, we see a lot of nice problems building from patterns to projects involving expressions. In new books and other materials, we see a shift in the way variables are addressed and introduced (but in many cases that shift has only been a pagination shift from pg. 55 to pg. 3). In projects, algebra seems to have lost some of its structural core and the cohesiveness that makes it a field of study as described by Hy Bass. In the texts, we still see hesitation to break free of the idea of a variable being an unknown and the relentless march from integers to quadratic equations via the solution of linear equations.

Algebra is beginning to be seen as the gatekeeper to quantitative literacy, in which quantitative literacy is viewed as comprising more than statistics. If we think of calculus as being either a pump or a filter, I think we have to look at algebra as being the faucet.

What do things like the NAEP results, the Conference Board of the Mathematical Sciences' (CBMS) report on undergraduate programs, and so on, tell us about the algebra enterprise? First of all, such information tells us that about 24 percent of our students today are taking a course called "algebra" in 8th grade or earlier. Further, we know that 94 percent of students have taken some sort of a course with "algebra" in its name by the time they reach the 12th grade. Of that 94 percent, one-half will have taken a second course in algebra called "algebra II" or something like "intermediate algebra" or "algebra/trigonometry." Beyond this point, we know that enrollment drops precipitously, with about 13 percent of all students going on to take a college preparatory mathematics class before high-school graduation.

Now, with that kind of a background, what do these students carry away from their algebra course or courses? I think the movie "Peggy Sue Got Married" captures it well. In it, actress Kathleen Turner is drifting back like a ghost into her high-school algebra class. Suddenly, she says that her algebra teacher lied because she never did need algebra in the world beyond that classroom. We teaching professionals—indeed most educated adults — have never successfully refuted her claim, at least not with students and not on a regular basis. This symposium provides us with an opportunity to focus on what school algebra is, how it can be structured, and to note the roles that algebra plays in the lives of citizens.

MODELS FOR UNDERSTANDING SCHOOL ALGEBRA

The 1988 NCTM Yearbook, *The Ideas of Algebra, K-12,* and the recently released *Approaches to Algebra: Perspectives for Research and Teaching* (Bednarz, Kiernan, Lee, 1996) point to several models we might consider in understanding school algebra. One of these is structure. How do we take a set of objects and impose an operation or a set of operations on them? How do we bring this beauty and structure to students and at what age? One might think of these objects as symbols. One might think of them as more concrete entities. With objects and operations, we can begin to look at properties and at statements of generality that we can put with them.

Much of the history of our doing this—algebra—has evolved from our actual work with computation and our ability to abstract from the structures surrounding computation. We all know about sequences of decreasing exponents on integers and using them to decide what a negative integral exponent means. We have taken a look at various patterns that result from operating with matrices, small matrices, two by two, and then trying to deduce from a set of these what an inverse might look like and if and when inverses might exist. We have developed properties, and, from these properties, we have created structures.

It is here that we begin to speak of algebra. Representations are developed for ideas that we have learned to manipulate verbally. There has been recent discussion about the rule of three: graphical, tabular, and symbolic representations. I think we have to have a rule of four, including verbal representation, and think of verbal and the other three representations as located at the vertices of a tetrahedron. The edges of the tetrahedron then serve as models of the shifts in representations that we hope students can engage in for a given concept.

How do we take all these words and translate them? How do we actually assess students' development of linguistic representations? How do we look at students' growth in making shifts along the edges of this tetrahedron? Is it the students' ability to make such shifts that shows us best the beginning of the students' depth of understanding?

Many concepts in algebra are tied up in language. Hy Bass just spoke about the geometric representations, but there are hierarchies within that representation. There is length, measured in centimeters. There is area, and it is represented in square centimeters. That is a bit more difficult to think about. Then there is volume, which comes in cubic centimeters. That is even more difficult for many students to conceptualize. And there are rates, which are comparisons of objects that have different units. Students can handle rates when doing comparisons of such things as price per soda can. However, if I start to talk about velocity or density, I hit a wall with students because these rates are not rooted in their concrete conceptions of rate. Part of the wall is the students' inability to think of objects, to tie them to a concrete model, and to develop the linguistic ability to function with that model. This is part of learning and thinking about algebra, an algebra of representations. There also is quantification: "each," "all," and "some." These words, along with "and," "or," and "not," play major roles in the translation of problems to the real world.

Another model for algebra is built around the study of functions and relations. There appears to be a great tension between the more structural, historical, polynomial-based approach and the functional representation approach. Some would say that the function approach is really more analysis than algebra. If so, where do we cross the line, and does it really matter? I think that these issues are not as important as developing a clear understanding by students of the mathematics that underlie what we want them to know and be able to do. This is especially true in situations involving simple systems and numerical relationships but also functions and operations that represent the relationships.

Finally, in terms of models, there is the modeling version of what algebra might be, which involves looking at and building models to represent real-world contexts and problems within those contexts. This approach often draws on statistics, analysis, geometry, and many other forms of representation.

What is most important as we take a look at algebra across the curriculum is for us to begin to think about the translations and the transitions that are necessary as students move from understanding arithmetic to beginning to understand divisive or rate-based structures and how to operate in them. Moving on to generalize examples in these situations is algebraic. This type of modeling approach relates back to the approach that builds relational models from the data.

Reflecting on the four representations of algebra that I just outlined—structural, linguistic, functional, and modeling—none of them can survive or prosper alone. In fact, what we really have to do is think of how to merge them to support a coherent program of study in algebra. That means developing a program of study that enables students to meet four goals of algebra in the context of the school mathematics program.

FOUR GOALS FOR SCHOOL ALGEBRA

The following four goals for school algebra programs are to help students see and use algebra as a way of

- representing quantity and relationships among quantities;
- predicting what happens in quantitative settings;
- controlling, where possible, the outcomes to quantitative processes;
- extending the applications and establishing the validity of new relationships in the structure of algebra.

These goals may look a bit different than the customary goals one would list for students of algebra. For example, we normally don't think about using algebra to predict, but that is exactly what we do when we build models and apply them. We use algebra in this fashion because it allows us to look at how the manipulation of objects could suggest what might happen. We use algebra to control when we establish a structure of statements that runs a computerized assembly line. We use algebra to control the paths of action we might want to follow in running a process.

Beyond these goals, we need to look at how an algebra program provides motivation for students; i.e., How does it address why they are studying algebra? How does it relate to their future and to their daily lives? How does it provide a basis for their continued learning? We can do many interesting things with algebra at several points in the K-14 curriculum, but what we do may not provide the foundation for the next steps students themselves may have to take or will want to take sometime in the future.

We constantly need to be looking at how we can provide a base for student growth in algebraic thinking. Along these lines, we also must work to prevent Peggy Sue's lack of connection between her algebra class and the outside world. I, too, have never had to package pecans, peanuts, and cashews to make mixtures in my life!

BIG IDEAS

There are a few big ideas that all students need to understand. The most important is variable. There are many roles for variables, and Zal Usiskin has listed and characterized them well in his lead article for the NCTM Yearbook (1988). Among the uses listed by Usiskin, I would place a great deal of stress on expression of functional relationships, when variable is a symbol expressing a quantity but, further, for expressing a quantity that varies.

From whence does algebra grow? It grows from the study of growth itself. One of the first places students see growth is when they look at patterns and patterns of numbers. When they arrange centimeter cubes to model a pattern of numbers such as 4, 6, 8, 10, 12, __, __, __,... , they are beginning to study growth. This pattern becomes apparent in something that they are doing with pattern blocks, such as putting them together and looking at the perimeters as they extend a pattern of consecutive squares. Or it may come as a result of a calculation. What is central, however, is the pattern of arithmetic growth.

Now, when we look at arithmetic growth at a later stage in our learning, we might look at this growth a little differently. We might look at it as an application of difference equations where the next value is the previous value plus some addition. In our pattern block example, four was my initial value, and then each next value was the previous value plus two; i.e., $p_{n+1} = p_n + 2$ and where $p_1 = 4$.

As elementary-school children begin to build out from their study of patterns, they are developing the foundation of the next big idea—*linearity*. Many children approach linearity recursively and then move to an understanding that approximates seeing linearity like a function of a real variable. However, this transition is really hidden from them, hidden, in fact, even from many undergraduate students. When we come to examine change in linear situations, we first see it as adding on unitary change, the idea behind slope. Eventually, we shift to developing algebraic formulas, written something like $p(n) = 2n + 4$, for the perimeter example we examined earlier.

The second form of growth students encounter is geometric growth. This is evident in situations having a constant multiplier. They may start with 2, and then each successive value comes from multiplying that previous value by 2. The result is the pattern of 2, 4, 8, 16, 32, 64,... . Another big idea is *exponential* growth. We want to be able to capitalize on student actions to develop other ways of looking at growth. Situations may have some additive parts and some multiplicative parts. Some situations may have some additive parts, some geometric parts, and some may even have a strange function tacked on that might be polynomial, logarithmic, exponential, trigonometric, or who knows what? We want students to be able to take a look at a growth situation, ferret out the patterns embedded in it, and develop algebraic expressions that allow them to predict, control, and understand what makes it tick.

This is one of the lessons that Peggy Sue missed. Algebra is the key to understanding change in the world about us. Consider the typical annual report graph from a Fortune 500 company. Such graphs commonly show how the Standard & Poor's index is moving, how the stock price of the company is moving, and how the volume of shares of the company sold in a given day has varied. Now, if you want to talk about things being a function of other things, all the little points along this graph are revealing of significant happenings, in the company or in the market itself, for example.

If one wanted to place a value on that company in order to sell it, all of these data points would be important for constructing a helpful mathematical model. Some of the models that investment bankers build are systems of difference equations. If you look at derivatives on Wall Street, you are looking at systems of partial differential equations. Mathematics and algebra play a very strong role in modeling financial things in our world. Peggy Sue, take note.

I want to step back into the student's world. Suppose you had the task of answering the following market research question. It is a common question involving which TV shows capture which portions of the audience:

Background
Consider the problem of market research associated with monitoring the audience share watching a given late night show. Each night the surveys show that 25 percent of yuppies watching the "Tonight Show" with Jay Leno watch it again the next night. Likewise, 50 percent of the yuppies watching the "Late Show" with David Letterman watch again the next night. The remaining audience shifts back and forth.

Problem
a) Model the number of individuals watching the shows over time with a system of difference equations.
b) Find the general solution for the system of difference equations.
c) Given that Tom Cruise and Cindy Crawford are on the season premiere of the "Tonight Show" and all yuppies watch that show, find the particular solution for the system of difference equations.
d) For the long term, determine what portion of the yuppies watch the "Late Show."

You can probably see the system of equations in the situation outlined above, but the system actually happens to be a system of difference equations. How would you solve it? Well, perhaps there is a relationship between how we would approach this and how we would approach a system of differential equations, equations that are algebraic in form and that also account for change.

In the past, we used systems of linear equations with algorithms like Cramer's Rule because they appeared to give us quick and efficient solutions. It was a paper-and-pencil method that was fairly efficient for very low-dimension systems. But the introduction of technology into the classroom has permitted us to start to use the power of algebraic structure to consider what we mean by efficiency. We quickly see that Cramer's Rule is not very efficient once the size of systems increases. Further, it requires us to add additional baggage to the school curriculum—determinants. In fact, once we start to take a look at solving systems of linear equations, we really want to take a look at what analogies we can build for our students. We can solve $ax = b$ as a polynomial equation $x = a^{-1}b$. But if we use the same type of structure, and we want to solve a system of equations, then we have coefficient matrix A multiplied by X, the variable matrix, equaling the constant matrix B. And this can be expressed by $AX = B$. Its solution, where it exists, is given by $X = A^{-1}B$. Are there some similarities here? What have we gained and what have we given up in making such changes in the content and approaches to our study of linear systems in a technological age?

Algebra is very much part of our world—as representations and as models. Our ability to take and pull structure from problems, our ability to look at the objects involved, and our ability to act on these objects as algebraic objects constitute the essence of what we are about in school algebra. How do we manipulate? How do we actually take and interpret the results of those manipulations? How do we look at the linguistic parts? How do we look at the relationships in these situations? Is this not the essence of what we are talking about at this symposium? And how we can make it happen for our students at all levels in grades K-14?

Today, the ways in which we represent things are changing. Bar codes and public key cryptography codes are now used to structure and represent number. And these structures have all types of uses. The numbers built into the ISBN codes on the back of a textbook or the grids of bars on your most recent FedEx package provide you with security. The digits on the back of the book you want have a check digit built into to them to make sure that the typist does not transpose them while entering your order in the computer. (The check digit is an algebraic property of such numbers making application of a structural property of numbers based on divisibility by 11.) A similar form of algebraic representation allows a can of Campbell™'s soup to be identified and accurately priced by the scanner in your local grocery store, as well as allows the grocer to make almost instant pricing changes. Bar codes are also used to identify individuals in security settings.

What I have tried to do is talk about some of the different models of how we represent algebra. How might we look at beginning to put them together in a program across grades K-14 for students? One of the first things we want to do is involve students with number and with data, and I think we want to do this at all levels. The variety of representations that number may take—from integers to matrices of complex numbers—will require a great deal of flexibility from students. They must learn to see that data can come in many forms.

We want students to look for and to begin to describe pattern within its context. Pattern within and among numbers calls on students to start to draw on their representational and linguistic skills. We want them to start to recognize the quantity that is embedded in the data they confront. What is that which can be measured? What is the quantity, that is, the essence of the particular setting? We might want students to move to expressions, to take that variable and be able then to put it into a form that makes use of and begins to look like what we call algebra. What we have done at this point—if not before— is to begin to solve equations. The essential difference is that we already have moved through the study of functional representation.

We have moved to function when we represent the pattern as $4 + 2x$, but we haven't carried all the trappings of functional notation with us. Interestingly, students are thinking functionally. They are looking at direct and joint variation. I think that we can look at an expression and that we can evaluate it; i.e., we can use formulas. I think students can begin to ask, does the form $2x + 4$ ever take on the value 100? If so, when? This leads to the building of real equations from expression forms to answer the question of whether a given number will ever appear in the list of potential outcomes. Finally, the student moves to modeling and building models. The sequence of steps involves rich data, recognition of pattern, developing quantity within that pattern, representing that quantity through the use of variable, developing a function-like expression to represent typical values within the pattern, using equations to study the pattern, and using the models developed to represent, manipulate, predict, and control with algebra.

How do we mold these visions of algebra into a coherent program for students in grades K to 14? How do we build the concepts of variable, table, graph, expression, function, equation, and others into a solid program—a program that deals with properties, principles, procedures, and proof and has the kinds of understandings I talked about earlier? How do we develop articulation in these programs? We no longer can afford high-school programs that don't accept 8th-grade programs or college programs that do not recognize good work done at the secondary school levels. We have to have school system stakeholders talking to one another across important articulation points. We have to know, to some degree, more about the foci at particular points in the curriculum because, without that kind of information about the curriculum, there is no way to monitor opportunities afforded to our children, our students, our future citizens. When no one has the responsibility for seeing that students have the opportunity to receive, experience, and move through quality instruction, or to build carefully a program that includes these important steps, we are in trouble. We have to think about how we assess progress in these important stages to learning and using algebra. This is essential because, in many cases in the past, assessment systems have been the biggest deterrents to changing school algebra.

Perhaps, at least, I have given you some food for thought about the history of school algebra and its status. The future is in our hands—in the hands of all who walk with and work with children in mathematical situations. I hope that the algebra programs we move to are more motivating for students—lead students to see the need for algebra and the dynamic roles it can play in their lives.

REFERENCES

Bednarz, N., Kieran, K., and Lee, L. (1996.) *Approaches to Algebra: Perspectives for Research and Teaching.* Dordrecht, The Netherlands: Kluwer Academic Publishers.

Coxford, A. F., and Shulte, A. P. (Eds.) (1988.) *The Ideas of Algebra, K-12.* 1988 Yearbook of the National Council of Teachers of Mathematics (NCTM). Reston, VA: NCTM.

Davies, C. (1846.) *Elements of Algebra.* New York: A. S. Barnes & Co.

Fullan, M. (1993.) *Change Forces: Probing the Depths of Educational Reform.* London: The Falmer Press.

National Council of Teachers of Mathematics (NCTM.) (1989.) *Curriculum and Evaluation Standards for School Mathematics.* Reston, VA: Author.

Presentations on Day One

Transforming Algebra from an Engine of Inequity to an Engine of Mathematical Power by "Algebrafying" the K-12 Curriculum (J. Kaput)

Developing a Coherent and Focused K-12 Algebra Curriculum (E. Phillips)

Enhancing Algebraic Reasoning with Technology (G. Akst)

Algebra for Everyone? With or Without Technology? (M. Norman)

How Might Technology Enhance Algebraic Reasoning? (R. Zbiek)

What Do We Know about K-14 Students' Learning of Algebra? (J. Confrey)

Algebra: What All Students Can Learn (S. Williams and D. Molina)

Improving K-14 Algebra Instruction: A Discussion of Teachers' Responsibilities and Students' Opportunities (B. Moore-Harris)

Transforming Algebra from an Engine of Inequity to an Engine of Mathematical Power by "Algebrafying" the K-12 Curriculum

James J. Kaput
University of Massachusetts
Dartmouth, Massachusetts

In this brief note I assume, first, that, just as algebra has acted as a constricted gateway to significant mathematics and all that follows from mastery of that mathematics, algebra *reform* is the gateway to K-12 *mathematics reform* for the next century; and, second, that by acknowledging the several different aspects of algebra and their roots in younger children's mathematical activity, a deeply reformed algebra is not only possible but very achievable within our current capacity for change.

The key to algebra reform is to integrate algebraic reasoning across all grades and all topics—to "algebrafy" school mathematics. This integration solves three major problems:

1. It opens curricular space for the 21st century mathematics desperately needed at the secondary level, space currently locked up by the 19th century high-school curriculum now in place across the nation;
2. It adds a new level of coherence, depth, and power to school mathematics, both as a curriculum and as a habit of mind; and
3. It eliminates the most pernicious curricular element of today's school mathematics— late, abrupt, isolated, and superficial high-school algebra courses.

An early-start, strands approach to algebra also fits well with an inclusive, big-idea strands oriented approach to the curriculum at large, which contrasts with the layer cake-filter structure that delays and ultimately denies access to powerful ideas for all students but a few. An algebrafied K-12 curriculum supports democratization of access to powerful ideas.

Our discussions of algebra must be as honest and clear as possible. Towards clarity, we need to distinguish "Algebra, the Institution" from "Algebra, the Web of Knowledge and Skill" that we want students to develop in school, so that criticisms of the former are not heard as statements about the latter. "Algebra, the Institution" is a peculiarly American enterprise embodying the standard courses, textbooks, tests, remediation industry, and associated economic arrangements, as well as the supporting intellectual and social infrastructure of course and workplace prerequisites, cultural expectations relating success in algebra to intellectual ability and academic promise, special interests, relations between levels of schooling, and so on. Exhortation for and legislation of "Algebra for All" tacitly assume the viability and legitimacy of this institution. But this algebra is the disease for which it purports to be the cure! It alienates even the nominally successful students from genuine mathematical experience, prevents real reform, and acts as an engine of inequity for an egregiously high number of students, especially those who are the less advantaged of our society.

Our challenge is to create an implementable alternative to this inimical institution, to transform an engine of inequity to an engine of mathematical power. One first step is to get a bit clearer about what we mean by "Algebra, the Web of Knowledge and Skill" so that we can see what it might offer as a goal to work towards and how it differs from the intellectual content of the courses that are part of institutional algebra. We can then ask how we might achieve this algebra of knowledge and skill.

FIVE FORMS OF ALGEBRAIC REASONING

In my view, algebraic reasoning is a complex composite of five interrelated forms of reasoning. The first two underlie all the others (kernels), the next two constitute topic strands, and the last reflects algebra as a web of languages. All five richly interact conceptually as well as in activity—to understand *this* algebra is to make connections. All five can and should be started early.

1. Algebra as Generalizing and Formalizing Patterns and Constraints, especially, but not exclusively, Algebra as Generalized Arithmetic Reasoning and Algebra as Generalized Quantitative Reasoning;
2. Algebra as Syntactically-Guided Manipulation of Formalisms;
3. Algebra as the Study of Structures and Systems Abstracted from Computations and Relations;
4. Algebra as the Study of Functions, Relations, and Joint Variation; and
5. Algebra as a Cluster of (a) Modeling and (b) Phenomena-Controlling Languages.

As kernels, forms 1. and 2. underlie all the others listed above, with the reasoning in 1. based both inside and outside of mathematics and the reasoning in 2. done in conjunction with 1. It is difficult to point to mathematical activity that does not involve generalizing and formalizing in a central way. It is one of the features of thinking that makes it mathematical. Also, the actions one performs with formalisms identified as 2. typically are not generalizing and formalizing per se but, rather, typically occur as the direct or indirect result of prior formalizing. The formalisms also may be of many different types, not merely variables over sets of familiar numbers (or transcendentals over some field). It is also possible for the manipulation to yield general patterns and structures at another level of generalizing and formalizing, which is the essence of 3., the structural form of algebraic reasoning. In order to use or communicate generalizations, one needs languages in which to express them, which leads to 5., and 5., in turn, permeates 1. through 4. While 3. is a school mathematics topic strand found today mainly at the advanced levels, it is also an important domain of mathematics in its own right— abstract algebra. On the other hand, topic strand 4.—functions—is more a school mathematics domain that lives in the world of mathematics more as a general purpose conceptual tool than as a branch of mathematics.

Traditional school algebra focuses on form 2. above at the expense of all the others. And while calls for a functions approach to algebra were ignored for almost a century, some of our contemporaries tend to say that form 4. is what algebra is all about. Although functions are an extremely powerful organizer of mathematical activity across topics and grade levels, this also is true of the other forms of algebra listed, which is exactly why algebra can play the key role across K-12 mathematics, as I and others have suggested. This wider view of algebra emphasizes its deep but varied connections with all of mathematics.

ALGEBRA BEFORE ACNE: THE ROLE OF TEACHERS

The language aspect of algebra supports both early and integrated algebra: early because students require repeated use of a language over an extended amount of time to become fluent in its use; integrated because in order to learn a language, students need to use it to express something significant to them, such as the quantitative relationships arising inside mathematics (for example, that occur in arithmetic and geometry) and outside mathematics (when we use it to model our world).

As several speakers will show during this symposium, appropriate instructional materials can "seed" each aspect of algebra that I have listed through relatively ordinary, elementary mathematical activity. One key is that teachers need to be able to identify and nurture these roots of algebraic reasoning in forms that appear very different from what is deemed "algebra" under the auspices of "Algebra, the Institution." For example, generalization initially can be expressed using ordinary language, intonation, and gesture rather than formal symbolism. This will require teacher development focused on student thinking rather than traditional formalisms. Another key is that these beginnings need to *go somewhere mathematically,* both in terms of growth in notational competence and in terms of the significance of the big ideas that these notations are used to express. This will require careful designed classroom materials to help guide the way.

Developing a Coherent and Focused K-12 Algebra Curriculum

Elizabeth Phillips
Michigan State University
East Lansing, Michigan

Successful completion of algebra is seen as a "gateway" to future success in scientific and other postsecondary endeavors. Getting through the algebra gate is so important that it has given rise to other gates, such as state-mandated algebra tests and a variety of pre-algebra courses that consist of a combination of arithmetic drills and simple equation solving.

The recent Third International Mathematics and Science Study and National Assessment of Educational Progress data (U.S. National Research Center, 1996; Mullis et al., 1991) confirm that the U.S. is falling short in the goal of improving the algebraic understanding of all students. One conclusion of the TIMSS curriculum analysis is that the American mathematics curriculum is a "mile wide and an inch deep." Such reports suggest a lack of coherence and focus in the K-12 mathematics curricula across the country, with little time devoted at any grade level to developing deeper understandings of fewer important ideas (Schmidt et. al., 1996).

Nowhere is this lack of focus and cohesion more noticeable than in the area of algebra. Continued demonstration of students' lack of basic skills despite two decades of a skill-and-drill curriculum should convince us to look for a new way of bringing students and algebra together. In the search for new ways of teaching and learning algebra, we must seek answers to the following questions: How do we build a coherent algebra curriculum across the grades that develops understanding of important algebra concepts and skills and students' quantitative reasoning abilities? How do we assess these skills and understandings?

The paper, "A Framework for Constructing a Vision of Algebra: A Discussion Document" (National Council of Teachers of Mathematics [NCTM], 1997), suggests that algebraic concepts are best learned first within a context or problem. While a contextual setting can become a vehicle for understanding and remembering these concepts, there is an intrinsic difficulty in building a coherent curriculum around a range of concrete contexts in which the student and teacher may not see any common patterns. The NCTM algebra paper offers a way to connect these contexts by using organizing themes: Functions and Relations, Modeling, Structure, and Language and Representations.

> "It is the ability to generalize, extend, apply, and connect ideas throughout many different situations and across several grades that enables students to make sense of mathematics; thinking of it as a set of unifying ideas rather than a set of disjointed, unrelated problems."
> "Algebra becomes manifest through exploration of the compelling ideas within the contextual settings, while the themes bring logic and organization to the discipline." (NCTM, 1997)

Several of the recent National Science Foundation-funded curriculum projects have organized their curriculum around interesting problems. If mathematical concepts are developed from problem situations or contexts, then the variables in the situation and relationship between them become real to the student. The problem context can provide the scaffolding needed to reason about the quantities in the situation and how they change in relationship to each other. Variables and patterns are central to a problem-centered curriculum; that is, what are the variables, and how are they related?

In the "Connected Mathematics Project" (CMP) (Lappan et al., 1995), one of the NSF-funded middle-school curriculum projects, we have found that, over time, through such experiences, students develop a rich language, moving freely among the different forms of representations, pictures, words, tables, graphs, and symbols. They come to understand what each form of representation can tell them about a situation, and they learn which representation best captures their reasoning about patterns they observe between related variables. They understand what kinds of questions or information are more easily explored through a particular representation of the data or a combination of representations. The understanding needed to move freely among different representations takes time to develop and hence is an important part of the curriculum throughout.

A problem-based curriculum, such as CMP, puts quantitative reasoning in the forefront and thereby provides the base on which students can investigate patterns of regularity among rates of change between the variables. It is the concept of "rate of change" that helps students identify, represent, and reason about families of functions, such as linear, quadratic, and exponential. Early on in the study of algebra, symbols are used to represent these situations, along with other representations. However, symbol manipulation, per se, is not the focus—modeling situations and understanding the functions used to build such models are the focus. The contexts help students intuitively use symbols to represent their ideas about the relevant variables and their relationships. Eventually, students work with symbols, free of context. They do so in investigating the general characteristics of a family of functions, including solving for specific values of examples that belong to the family. Additionally, they investigate the syntax of the language of symbols as they look more closely at ways to represent problem situations symbolically, particularly those that give rise to different but equivalent expressions (Phillips and Lappan, 1997). The CMP curriculum suggests that algebraic reasoning can be developed by exploring a variety of contextual settings connected by organizing themes: the problems bring meaning to the themes.

More importantly, the students in the CMP curriculum are developing "habits of mind." As they move through the grades, the following questions become an integral part of their algebraic reasoning:

> "What are the variables? What are some patterns that relate these variables? How are these patterns captured? Can you predict from these representations? How are the variables changing in relation to each other? How is an increase in the independent variable related to a change in the dependent variable? Where is the dependent value changing the most? The least? How can this change be seen in a table? Detected in a story? Observed in a graph? Read from symbolic representations? Where does the graph cross the x-axis and the y-axis, and what is the significance of each of these intersections? Where does the dependent value reach its greatest value and its least value, and what is the significance of each? What effects does a change in parameter have on the situation?" (CMP) (Lappan et al., 1995)

Taken altogether, the intuitive uses of algebraic thinking in a problem-based curriculum plus the specific development of the concepts of variable and relationships among variables constitute a powerful, sense-making approach to algebra in the K-12 curriculum for all students. However, these practices are a radical departure from the traditional algebra curriculum that has focused almost exclusively on manipulating expressions and solving symbolic equations.

These new curricula put new demands on teachers: to teach mathematics through rich problems requires the teacher to have a deeper understanding and a broader view of mathematics and a deeper knowledge of a pedagogy based on inquiry. Teachers for the K through 12th grades are key players in the improvement of student learning of algebra. Any effort to reform the algebra curriculum or methods of teaching must help them to see the potential for developing mathematical power for their students and help them communicate this mathematics power to parents. The keys are student growth and performance.

QUESTIONS FOR DISCUSSION

What do you see as the issues or big questions concerning the development of

- a coherent algebra curriculum across the grades K-12;
- criteria or lenses for selecting curriculum, appropriate pedagogy, and assessment; and
- long-term support for teachers and *Standards*-based algebra?

Also, where do we go from here?

REFERENCES

Lappan, G., Fey, J., Friel, S., Fitzgerald, W., and Phillips, E. (1995.) *The Connected Mathematics Project.* Palo Alto, CA: Dale Seymour Publications.

Mullis , I., Dossey, J., Owen, E., and Phillips, G. (1991.) *The State of Mathematics Achievement: NAEP's 1990 Assessment of the Nation and the Trial Assessment of the States.* Washington, DC: National Center for Educational Statistics.

National Council of Teachers of Mathematics Algebra Working Group. (1997.) "A Framework for Constructing a a Vision of Algebra: A Discussion Document." Final report to the Board of Directors. East Lansing: Michigan State University.

Phillips, E., and Lappan, G. (1998.) "Algebra: The First Gate," in *Mathematics in the Middle,* edited by Larry Leutzinger. Reston, VA: NCTM.

Schmidt, W., et al. (1996.) *Characterizing Pedagogical Flow: An Investigation of Mathematics and Science Teaching in Six Countries* (pp. 49). The Netherlands: Kluwer Academic Publishers.

U.S. National Research Center. (1996.) "TIMSS" (Report No. 7) (pp. 9). Washington, DC: National Center for Education Statistics.

Enhancing Algebraic Reasoning with Technology

Geoffrey Akst
Borough of Manhattan Community College/CUNY
New York, New York

We tell our students that mathematical questions should be well defined. In mathematics education, however, the important questions are usually ambiguous.

Consider, for example, the question of whether technology can be used to enhance algebraic reasoning. Our response must depend on which technology we have in mind and on what meaning we attach to the term "algebraic reasoning."

To start with the latter point, suppose that we construe algebraic reasoning narrowly: suppose we say it is deducing consequences from a symbolic relationship. Solving an equation, a kind of proof, is a case in point. Technology— even the humble four-function calculator —can help to foster this kind of reasoning by relieving the student of computational distractions and complexities, allowing him or her to focus attention on the structure of the algebraic argument. And a graphing or symbol-manipulating calculator can be used, at a minimum, in checking the validity of the conclusions drawn, step by step.

Alternatively, suppose that we broaden our conception of algebraic reasoning to include a kind of intuition. For a student with this "algebra sense," symbol manipulation is a meaningful activity, and unreasonable answers are apparent. In this interpretation, algebraic reasoning includes the ability, through pattern recognition, to conjecture, even to have confidence in, symbolic generalizations. Surely technology has a role to play in developing this algebra sense. A student can use a computer algebra system to run through many examples of polynomial multiplication and develop the ability to predict the degree of a product or the number of terms the product contains.

Finally, the case can be made that algebraic reasoning implies the ability to employ symbolic techniques in attacking word problems. Even anti-calculator educators must admit that the more realistic and less restrictive applications which calculators and computers allow beginning students to tackle can motivate interest and, hopefully, increase time-on-task.

No matter how it is interpreted, algebraic reasoning can benefit from technology, even when we put technological tools aside. High-tech instructional delivery systems of increasing sophistication now present students with motivating lessons. CD-ROM-based computer programs and the World Wide Web are increasingly becoming attractive environments in which to learn about all subjects, including algebraic reasoning.

Algebra for Everyone? With or Without Technology?

Milton O. Norman
Granby High School
Norfolk, Virginia

Americans have placed an increasing number of unusual burdens on their educational system. Schools are expected to teach basic skills in reading, writing, mathematics, and reasoning. They also are supposed to introduce students to a complex cultural tradition in the liberal arts and the sciences, a tradition that now includes the entire world, not simply Western civilization. They further are charged with developing individual skills relating to such nonacademic areas as vocation, leisure-time activity, and citizenship, as well as providing such specialized programs as driver education, sex education, drug education, and parent education. As social institutions, schools are expected to be, simultaneously, agents of social stability and social change.

By the 1970s, apprehensions grew that American schools were not accomplishing all these objectives: indeed, the objectives themselves were questioned as possibly unreasonable. The back-to-basics movement at the end of the 1970s was an attempt by educators to narrow and sharpen their ambitions. Other efforts began in the 1980s to restore interest in academic quality and excellence. *A Nation at Risk*, a 1983 report issued by a bipartisan federal commission, emphasized the need to upgrade American education at all levels.

Technology is seen as a solution to upgrading mathematics education in general and to making algebra accessible to all students in particular. In simplest terms, the concern of the student of mathematics is "why" and of technology, "how." The relationship between the two is actually much more complex, however; some technologies are mathematics and science intensive, whereas the manufacture of such items as cigarettes or furniture depends much less on mathematics and science. Education must harness the particular skills of technologies to introduce in a more effective way concepts that were previously reserved for a selected group of students. Technology within the classroom has become a large issue, however, because of the many inequities in the United States. Some schools are without graphing calculators; others have full Internet access. This type of issue cannot be addressed solely by documents that propose the use of appropriate methods of teaching higher level skills. Rather, they must be addressed by adequate funding for equipment and essential training. Classroom teachers do not have the innate ability to grasp all the new technology introduced every day.

How should we teach algebra? Over half of the objectives in the current "Standards of Learning" within the state of Virginia suggest the use of the graphing calculator and/or computers to reinforce algebraic concepts. The argument for the graphing calculator is its accessibility to each student and its mobility. The handheld computer/ calculator is capable of complex Computer Algebra Systems (CAS).

The technologies appropriate for mathematics depend on an individual teacher's training, experience, and the information disseminated by forums such as this one. We must be cognizant of the differences in experience, philosophies, and political climate within individual states and school districts. These are all major contributors to how we, as teachers, will prepare students to participate in the new age.

How Might Technology Enhance Algebraic Reasoning?

Rose Mary Zbiek
The University of Iowa
Iowa City, Iowa

What is algebraic reasoning? The collective "Mathematics as Reasoning" standards (National Council of Teachers of Mathematics, *Curriculum and Evaluation Standards for School Mathematics*,1989) call for students to justify their answers and solutions, to make and evaluate mathematical conjectures, to use counter examples effectively, to draw logical conclusions through deductive and inductive reasoning, and to appreciate the role and power of reasoning in mathematics. Some combination of these goals is appropriate regardless of how we define algebra (e.g., generalized arithmetic, study of patterns, study of structure, language, study of symbolic manipulations, problem-solving tool). For our purposes, algebraic reasoning includes engaging in these activities with planned or unplanned use of, or conclusions about, properties of, uses of, and operations on algebraic entities (e.g., variable, equation, function).

How can technology make a difference? Technology in mathematics classrooms takes many forms: mathematics tools (e.g., spreadsheets, dynamic geometry packages, computer algebra systems), special purpose programs, simulations, and communication tools (e.g., e-mail, presentation generators). The promise of technology in algebraic reasoning is not to do the reasoning for students but rather to provide a tool that facilitates reasoning, an environment that necessitates reasoning, and a means for communicating reasoning. Technology in its various forms has particular promise in several areas of algebraic reasoning, as follows.

Symbolic reasoning. Computer algebra systems (CASs) challenge the traditional algebra curriculum designed to teach students the skills that the CASs can do rapidly and relatively accurately. Students need to know the conditions under which the skills or algorithms apply. They also need the algebraic equivalent of numeric estimation and prediction, which includes knowing what form the result will take (e.g., number, expression). Similarly, they need to know what information equivalent forms yield.

Graphic reasoning. In the classroom where technology is present, students can develop understandings about graphs that parallel their understandings about symbols. They need to be aware of the ways in which we can operate on and with graphs. This graphic thread extends to include scatter plots and other graphic forms.

Multiple-representation reasoning. Technology today not only offers students a choice of representations but also links these representations. In the most common scenario, student-made changes in a symbolic form cause changes in graphic forms and tabular forms. Other scenarios might be geometric figures linked with data displays. Thinking about what happens and why it happens can lead to richer, more flexible understanding of the mathematical constructs embodied in those representations. In solving problems and studying other mathematics, students may move more easily across representational boundaries to solve problems and to develop understandings.

Structural reasoning. Technology can help students experience mathematical structure in ways that extend and connect symbolic, graphic, and numerical reasoning. For example, students can explore closure by analyzing multiple examples and verifying a general case of what happens when we use a binary operation with two objects (e.g., whether adding two linear functions yield a linear function). Similarly, they can consider inverses and other properties that underlie algebraic work.

Symbol-system reasoning. No matter what view of algebra we emphasize and what form of technology we use, we invariably have students reasoning within symbol systems. Time previously spent dwelling on manipulating expressions, equations, functions, matrices, and systems can be allocated to studying these things as mathematical entities that are related to other objects within these systems.

Integrated reasoning. Curricula that integrate rather than separate areas of mathematics and integrated technology allow students to move among these various areas. For example, data collected in a dynamic geometry environment can be displayed as a scatter plot, and subsequent discussion of fitted functions can lead to better understanding of both algebraic and geometric constructs.

Inductive reasoning. The speed and pizzazz with which technology can help us quickly to generate, store, and communicate multiple examples can be intoxicating. Conclusions based on data are likely to be no better than the data on which they are based. Students need to reason about the data and the collection as well as the patterns.

Modeling reasoning. Mathematical modeling transcends applying known mathematics to well-defined, perhaps contrived, contexts. Students need to move and reason flexibly between real-world experience and mathematical constructs.

The incorporation of technology in the algebra curriculum provides opportunities for students to ask "what if" questions, to pursue "how come" questions, and sometimes to face "what happened" questions. There are many questions, however, about how to use this to facilitate students' algebraic reasoning. These questions include, What are the "basic skills" of algebraic reasoning? What combination(s) of learning tasks and technology best facilitate long-term understanding as well as present-time reasoning?

REFERENCE

National Council of Teachers of Mathematics (NCTM). (1989.) *Curriculum and Evaluation Standards for School Mathematics.* Reston, VA: Author.

What Do We Know about K-14 Students' Learning of Algebra?

Jere Confrey
Cornell University
Ithaca, New York

In far too many school districts in the United States, we see intolerably high failure rates in algebra. In 1992-3, the Indicator Reports from the National Urban Schools reported an overall failure rate of 65.5% in a first course in algebra. Well over half of the students enrolled in the course were failing. Considering that these data were collected prior to the "algebra for all" movement, one might predict that if no changes are made in the practices of teaching algebra, these failure rates could increase as more students enroll in the courses. (Note that this prediction is belied by data from New York City, where a change in policy to fuller inclusion led to no decrease in the success rate.) Nonetheless, it is fair to say that we are in a serious crisis. This crisis deepens as one considers the desegregated demographics indicating that the failure rates for Hispanics and African Americans are at the high end of the scale (65% and 71%) in comparison with failure rates for whites (36.5%) and as one notes that the first two populations are projected to increase significantly in percentage of the total population. To find strategies that lead to successful entry to higher mathematics for each of the population groups in this country, whether these groups are distinguished by age, size, language, ethnicity, gender, and/or class, is indeed a national imperative. Being willing to challenge and change the content, curriculum, pedagogy, and assessment is an essential part of this process.

What is it in the research on student learning that can inform these changes? First of all, the data show that with enough resources, quality instruction, and time on task, anyone can learn introductory algebra. However, in our places of most need, few of these resources, in and out of school, human and material, are readily available. Thus, we need to consider what we know about successful learning of algebra and strategically and aggressively implement those approaches and findings. I will summarize the research under four headings: the importance of a functional approach with contextual problems, the use of multiple representations, the importance of student methods and interactions, and the systemic or community approach to change.

A FUNCTIONAL AND CONTEXTUAL APPROACH

Most textbooks still begin algebra with solving for unknowns. Treating algebra as a "generalized arithmetic," teachers expect students to learn to decode complex syntax, and, typically, to solve for *x*. In these settings "*x*" is just an unknown, and its solution is either right (matches the teacher's or solves the equation), or it is wrong. A great deal of research show that even when students learn to carry out these sequences faultlessly, seldom can they adequately explain or justify their solutions. Implicit grammatical rules trip them up, and algebraic manipulation is only learned by those who, through diligence, repetition, and practice, are willing to gain familiarity and fluency in the rules and procedures. Those who flourish when playing mental games, who find languages intrinsically interesting, or who spontaneously respond to puzzles find the endeavor worthwhile and rewarding. For most, what is learned is too often promptly forgotten.

Research on symbol manipulation predominantly has consisted of a documentation of error patterns. To know common errors is useful to teachers. They can try to flag the problem areas, draw students' attention to the correct procedures, and provide controlled practice to encourage progressive automaticity. Unfortunately, the data show that these "fixes" often fail over short periods of time, and the errors resurface. Improvement is incremental and modest. The loss of student participation, approximately 50% per year, far exceeds the modest instructional gains.

Other research has been devoted to designing a functions-based, context-based approach. A functions-based approach means that, instead of first teaching a whole layer of algebra as consisting of methods for solving equations and working with unknowns, one uses from the beginning the idea of relationships between or among quantities or sets as the basis for algebraic thinking. The context-based part of the approach means that the algebraic activities emerge from problems and tasks that relate frequently—but not necessarily exclusively—to referents from everyday activity. Researchers have offered a multitude of arguments for this, including the ease of identifying situations to which students can relate; the value of motivating math with a need to explain, predict, and model and thus actively to involve students in useful mathematics; the value of position, time, and rate to make the idea of variation accessible, to anticipate calculus, and to lessen dependence on symbol strings and to vary representations with graphs and tables, for example; and how making algebraic manipulation skills instrumentally rather than intrinsically valuable and using new technologies to carry out the algorithms of traditional algebra allows students to focus on using problem solving, communication, and learning technologies to engage in rich modeling-based activities. Focusing on functions could have some limitations. These could include a tendency to eliminate the study of the simple, elegant structures inherent in finite algebras, a lack of attention paid to proof and theorem-building, and inadequate time devoted to developing fluency in algebraic manipulation skills.

Just as not all of the benefits listed can be attained by any presentation of a functional approach, neither will all of the potential limitations apply. Teachers or curriculum writers can choose what to emphasize and de-emphasize. In my research on functions-based approaches in 7th grade, 11th grade, and in university-level precalculus courses, I have found that, within contextual situations, an effective and knowledgeable instructor can carefully and intentionally incorporate and highlight the issues of focusing on structure, proof, and fluency in discussions and assignments. When this is done, student reasoning is strong and well-grounded. For instance, in a two-year study involving an eight-week introduction to algebra to an entire 7th grade class using functions, computers, and motion detectors, one teacher commented,

> "In the old, standard way of doing algebra, everybody has to do it in this strict format, and there's no creativity or freedom of self-expression, no individuality to it. So you lose a lot of kids because they don't want to get boxed in or they want to be able to do it their own way. So they try and create it their own way, and it falls apart on them, and there's nothing for them to fall back on. There's no support, you know, 'I think *this* should happen' and if that doesn't happen, boom, you're off the cliff and that's the end."

In contrast, after using contextual problems and a functional approach, the same teacher commented,

> "When they get to solving an equation, their explanations are not, 'I need to subtract because it cancels multiplying.' It's, 'I need to subtract, because I need to get this hiker back to where he started.' They're using situations that they've seen along the way to explain things that they used not to have explanations for beyond, '...that cancels that out.' [Students can] believe their mathematical reasoning is right rather than just thinking it's right because 'I followed the right steps.' "

Research on a functional approach with contextual problems has shown that students tend to prefer this approach. They find the use of "real-world" problems to be highly motivating, and they show evidence of deeper thought at younger grades. Instead of focusing exclusively on the manipulation of symbols, they can move from problem context to problem context, often using these as the comparative structure rather than the symbolic algebraic form. The challenge becomes how to order the context so as to be assured that the students are gaining progressive conceptual development and also that their use of such powerful tools as graphing calculators and computer software is accompanied by a flexible and deep level of understanding.

At the same time, our research has shown that teachers find it challenging to move to a functional and contextual approach. Many of them were successful in traditional settings and lack experience in problem solving. Furthermore, unless a functional or contextual approach is aligned with methods of assessment, teachers tend to minimize these activities and to steer towards easily accessible and predictable skill development. Their use of new technologies is all too often limited to simple inductive inferences from a few examples or to a display of the graphical image and identification of its critical points (vertices, intercepts, symmetries, etc.). For teachers to use new technologies to explore new content opportunities or as everyday or practical tools takes extensive professional development.

THE DEEP USE AND COORDINATION OF MULTIPLE REPRESENTATIONS

Pick up a reform document and you will see a call for the use of multiple representations. However, in most treatments, we still see the obvious preference for the symbolic equation over any other type of representation. This occurs in part due to technologies that are heavily symbolically driven or limited in screen space. In textbooks, lip service is given to the rule of three (graphs, tables, and equations), but the internal logic of working with each representation is given short shrift.

In contrast, researchers have documented the potential of students to work actively and extensively in a variety of representations. This includes both the use and the invention of the representation. In the use of data tables, functional approaches often lead to the merging of statistical ideas with algebraic approaches. Similarly, in graph-related activities, tools such as "Function Probe" and even geometric sketchpads can lead to an integration of the visual, analytic, and symbolic. In our research, we have documented the essential role of the table in the development of functional reasoning and notation. Once the symbolism was available, the use of the table declined, but today, spread sheet capabilities could help us think about how to use tables once again. Graphically, we have shown that there are multiple approaches to the topic of transformations and that the "best" method depends on one's preference for a $f(x)$ notation versus a more symmetrical x-y form of notation.

In other work, researchers have documented the potential of introducing graph shape as a means to gain facility in the use of notation. Graph shape tends to highlight issues of rate of change and, with careful and early introduction, issues of accumulation. For example, students can view an exponential function as a bar graph and see the area under the curve as the sum of the previous bars. Hearing again from a classroom teacher, we report,

> "We looked at graphs qualitatively [in a college physics class], and it didn't make sense to me then. And it made more sense to me looking at it the [way] we did [in this project]. Think about letting kids look at graphs without having specific data, that was new. Just look at general forms of graphs rather than having particular pieces of data. That wasn't something I'd ever thought of doing."

STUDENT METHODS AND INTERACTIONS TO GENERATE MATHEMATICAL INVENTION AND DISCUSSION

Critical to understanding a reform approach to algebra is considering how it fits into the larger framework for reform, including creating a need for the ideas and subsequently developing formal description as a means to communicate and make connections. As documented repeatedly, there are verbal challenges when the formal meanings are merged with informal ones. The solution of obliterating or erasing the informal meanings seldom works. More effective seems to be uncovering the rational path that was taken to evolve towards the formal definition. Also critical is a thorough understanding of the possibility of alternative definition. Learning to reach a consensual decision—for clear communication—is important.

Reform approaches to algebra reach deep into our assumptions about learning. They challenge us to question our beliefs about who can learn and seek to eradicate harmful stereotyping. The use of rich contexts and tools and interactive and collaborative groups leads us to question the validity of theories of learning that only value individual progress and ignore group dynamics. Research on these contexts reminds us also to be wary of the idea that if all students participate in groups, all leave equally prepared.

Finally, research in the use of new technologies and project-based learning allows us to question many peoples' assumptions about the required developmental sequences. It appears that when students operate in complex settings with powerful tools, the sequences of learning vary, and the insights that are gained are more a function of the tasks and tools than any rigid developmental sequence. Paramount among these are assumptions about the description of concrete versus abstract thinking.

SYSTEMIC AND COMMUNITY CHANGE TO SUPPORT HIGHER ALGEBRA SUCCESS

Success in algebra is not simply a matter of getting individual students to develop psychologically or cognitively. The course acts within the community of learners as a signal of students' academic acuity and high career aspirations and expectations. Some students are provided extensive help on it at home, while other lack those resources and must either "get it" at school or cope with failure or exclusion. Since algebra is a gateway to advanced math course taking, its importance in the curriculum exceeds many other uniformly accessible courses or electives. For this reason, participation and success in it has been cast by some researchers as an issue of civil rights. It is at this critical juncture that our decisions about content, sequence, pace, curricular approach, use of technology, forms of interaction, types of help, and feedback to students carry out a political agenda that has ramifications way down the line in terms of participation and success.

Research has shown that community and parental involvement can be critical to students' persistence and success. It also has shown that reform is difficult for teachers on their own and that the systems can respond to change with institutionalized racism, classism, and sexism. Even as females succeed at equal rates, their less favorable attitudes about the domain and the tendency of their male peers to sex-stereotype make this an important territory for intervention and counseling beyond discussions of cognitive approaches.

In conclusion, research on reform in algebra suggests that progress will depend on a systemic response. Teachers must be deeply and consistently involved, must guide the process, and must be held accountable for equitable outcomes. We must introduce early the basic ideas so all children are well prepared for the transition and use a variety of materials and technologies, address the deep content needs of teachers in implementing reform, involve the larger communities, and constantly monitor our progress toward the important outcome of high achievement in algebra as a equal opportunity for all students.

Algebra: What All Students Can Learn

Susan E. Williams
Department of Curriculum and Instruction
The University of Houston
Houston, Texas

David Molina
The Charles A. Dana Center
The University of Texas at Austin
Austin, Texas

It is common for people to believe that they never use algebra after their formal course work and, hence, to believe that algebra is not important. The truth is, they may not use algebra in the way they experienced it in school, but algebra is prevalent in their lives. Algebra has been used as a vehicle for developing important skills necessary for further work in mathematics, but in education, the power of algebra—its application in everyday situations—has seldom been acknowledged.

WHY IS ALGEBRA IMPORTANT?

Traditionally, algebra has been approached as a generalized structure of our numeration system. Students have been asked to solve problems that involve rote computations in sterile or contrived settings. According to the Board of Directors of the National Council of Teachers of Mathematics (NCTM Board, 1993), the algebra of the past advances only a narrow range of by-hand skills for transforming, simplifying, and solving equations—most divorced from any natural context. The algebraic understandings cultivated in this way are far removed from those needed in a technological world by both employment-bound and college-bound students.

The algebra that most people use on a regular basis consists primarily of generalized and specific examples of function relationships that illustrate the systematic dependence of one quantity on another. This perspective of algebra is different from the perspective evident in traditional school algebra, but it is equally important. Whether one realizes it or not, determining gas mileage, predicting the amount of food to prepare for a party, and figuring the costs of renting videos are all examples of daily situations steeped in algebra and requiring a certain level of algebraic understanding.

So is algebra for everyone? The answer depends in part on how one perceives algebra. Not every student needs proficiency in symbol manipulation skills. By choice or circumstance, many students will never reach the levels of mathematics study where they will use these skills. However, every student needs to understand how quantities depend on one another, how a change in one quantity affects the other, and how to make decisions based on these relationships.

Every student is algebraically educable, although not every student needs to know how to simplify rational expressions or to derive the quadratic formula. Every student is capable of (a) learning about the use of symbols, (b) using patterns to look for generalizations, and (c) understanding the use of dependent, systematic relationships to model situations and make predictions.

WHY IS IT IMPORTANT FOR STUDENTS TO BE SUCCESSFUL IN ALGEBRA?

The need for knowing algebra may not be as obvious as the need for knowing how to read, write, and do arithmetic, but the lack of such knowledge blocks the entrance to more advanced academic and vocational studies and often limits job advancement opportunities.

Successful completion of high-school algebra appears to open the doors of the mind as well as the doors of opportunity. A recent study by the College Board (Hawkins, 1993) revealed that more than 70% of the students who took one or more years of algebra attended college within four years of high-school graduation. High-school students with one year or more of algebra were more than twice as likely to attend college as students who did not take algebra.

The College Board also reported that, in general, majority students who complete high school algebra attend college at a much higher rate than underrepresented minority students. For example, at the national level in 1993, 58% of all white students as compared to only 47% of all African American students attended college. But successful completion of high-school algebra and/or geometry virtually eliminated the gap. For students who completed high-school geometry, 83% of the white students compared to 80% of the African American students enrolled in college.

This description of the College Board's perspective of "Algebra as the Gatekeeper" raises concerns not only for educators but for the nation as a whole. The business community has expressed its concern by demanding that high-school graduates be given access to the entire mathematics curriculum (in particular, algebra, geometry, and calculus concepts). Important aspects of the curriculum have been withheld from students who have not mastered the basic skills of mathematics. Students today need more than basic skills; they need technical skills (including algebraic reasoning) to obtain and retain jobs. In the past, technical skills have separated the "cans" from the "cannots," but in the future, these skills will create an even greater division in our work force. Those with technical skills will be more in demand and, as a result, will earn higher wages. The "cans versus cannots" will be replaced with the "haves versus have nots."

Students of all races have been guaranteed equal education, but the students who most often are tracked into low-level mathematics courses are minorities and/or those of low socio-economic status. Data from the National Assessment of Educational Progress (NAEP) indicates that less than 50% of the students in urban schools take any mathematics beyond one year of algebra, and 20% never study algebra at all (Mullis et al., 1991). This disturbing situation has led Bob Moses, a former mathematics educator and now noted civil rights activist, to identify access to algebra as an issue for a new civil rights movement (Jetter, 1993; Michelmore, 1995).

HOW DO WE TEACH "ALGEBRA FOR ALL"?

Whether the goal is access or increased mathematics achievement, finding the "Right Algebra for All" is a goal that we need to embrace (Chambers, 1994). We know from experience that the current school approach to algebra is too abstract and an unmitigated disaster for most students (Steen, 1992). This rigorous approach has served a purpose—one of identifying groups of students who have proven worthy of continued study in the sciences. This may not have been the intended purpose; nonetheless, it has and continues to separate and "track" our students.

If we are to develop an algebra that can provide all students with tools that will allow them to advance in both academic and employment settings, then administrators, curriculum leaders, and teachers are needed to help guard against "Algebra for All" as being interpreted to mean that every student should be placed in what one might call a traditional first-year algebra class. This approach will simply allow many students to experience failure at a higher level, according to Jack Price, NCTM Past President (NCTM Board, 1993). If "Algebra for All" is to become a reality, it is vital that algebra's connection to the world and its usefulness be evident. Students must be allowed to recognize the role and place of algebra.

We need to use the information that we now have about how students learn best and apply that knowledge to our teaching. Students need to be involved in active learning situations, with new information being connected to the information they already possess. Algebra students need to be provided with opportunities to experiment and collect data, to analyze that data, and to draw reasonable conclusions based on the findings. Algebra needs to focus on "ways to solve a problem" rather than "the way to solve a problem" so as to equip students with multiple solution strategies. A major obstacle in creating such courses stems from the fact that teachers have never been taught how to teach in a way that engages a diverse student population in a variety of interesting and important learning activities structured to develop the inherent critical thinking and problem-solving abilities of all students (Seeley, 1993).

If administrators pressure teachers to make algebra accessible to all without allowing for planning and development time for teachers, the outcome will lead at best to mediocre results and has the potential to subject students to even worse situations than they are already experiencing. Teachers who are convinced that the goal is not attainable or who are uncertain of a viable approach may either (a) subject students to the rigors of traditional algebra and confirm their beliefs that many students just can't do algebra or (b) water down the content of their traditional course so that *all* students can succeed. Neither massive failure nor grade inflation are the desired outcomes of the algebra reform movement.

To make "Algebra for All" a reality means focusing on fewer topics organized around central ideas and preserving mathematical integrity and rigor through the depth of study rather than the breadth of study. If we face reality, our students today have different needs for their high-school mathematics courses, and the courses must now serve different purposes. According to Cathy Seeley of the Texas State Systemic Initiative, it is not that we must lower the level of what we teach so that all kids can do it nor is it that we must simply raise our expectations and do the same thing for all students that we have done for some. Rather, we must shift the content and approach of what we do in order to come closer to serving some purpose in our students' lives.

Given a school with teachers eager to meet the challenge of creating an "Algebra for All" and armed with a vision for implementation, the probability of sustained change is dim without the support and encouragement of the school's campus leaders. History has proven over and over again that strong campus leaders are the heart of successful programs. We ask campus leaders to join mathematics educators in rethinking the purpose of algebra, in verbalizing the importance of algebra for every student, and in supporting efforts to create such a program.

Counting on You: Actions Supporting Mathematics Teaching Standards (National Academy Press, 1991), a document created by the National Research Council's Mathematical Sciences Education Board, provides an action plan for school boards and school administrators to revitalize mathematics education by the year 2000. The seven key action items in the document are as follows:

1. Establish mathematics goals and programs in your schools that are consistent with national (and state) standards for curriculum and evaluation.
2. Provide sufficient resources (equipment, time, budget) to support an instructional program meeting the standards.
3. Discontinue use of standardized tests that are misaligned with national (state and local) standards for curriculum.
4. Provide time for your teachers: time to plan and evaluate their own teaching, consult with colleagues about teaching, and confer with supervisors about teaching effectiveness.
5. Give your teachers freedom to exercise their ability, judgment, and authority and involve them in decision making.
6. Institute a comprehensive inservice program consistent with national (and state) standards and involve teachers in its development and planning.
7. Evaluate teachers by using information gathered from various sources—the teacher's goals and plans, students' accomplishments, repeated classroom observation—and national professional standards.

CLOSING REMARKS

Mathematics teachers need to understand algebra very differently than we did a few years ago. Success in algebra in the past has been judged by mastery of a set of manipulative skills presented in isolation. In contrast, the algebra of today is expected to eliminate artificial and meaningless exercises and to de-emphasize symbolic manipulation while at the same time giving students opportunities for exploration and for conjecturing and discovering. Not only must we understand algebra differently, we must be prepared to illuminate others. The algebra reform movement put into motion by the NCTM *Curriculum and Evaluation Standards for School Mathematics* (1989) will not succeed if teachers are not prepared to rethink (a) what is algebra, (b) how do we teach algebra, and (c) who can be successful in algebra.

It is actions such as those listed in this paper that will provide teachers with the initiative and courage to experiment with new instructional strategies and materials, to collaborate in their planning, and to recognize ways to attain the goals of the school and the district.

REFERENCES

Chambers, D.L. (1994.) "The Right Algebra for All." *Educational Leadership*, 51(6), 85-86.

Hawkins, B.D. (1993.) "Math—The Great Equalizer: Equity 2000 and QUASAR, Improving Minority Standing in Gatekeeper Courses." *Black Issues in Higher Education*, 10(6), 38-41.

Jetter, A. (February 21, 1993.) "Mississippi Learning." *The New York Times Magazine*; 28-32; 50-51, 64, 72.

Michelmore, P. (1995.) "Bob Moses's Crusade." *Reader's Digest*, 875(3), 107-111.

Mullis, I.V.S., Dossey, J.A., Owen, E.H., and Phillips, G.W. (1991.) *The State of Mathematics Achievement: NAEP's 1990 Assessment of the Nation and the Trial Assessment of the States.* Washington, DC: National Center for Education Statistics.

National Council of Teachers of Mathematics. (1989.) *Curriculum and Evaluation Standards for School Mathematics.* Reston, VA: Author.

National Council of Teachers of Mathematics Board. (1993.) "Board Approves Statement on Algebra." *NCTM News Bulletin,* 30(6), 1, 3, 6.

National Research Council, Mathematical Sciences Education Board. (1991.) *Counting on You: Actions Supporting Mathematics Teaching Standards.* Washington, DC: National Academy Press.

Persell, C., and Cookson, P. (1982.) "The Effect of Principals in Action" in *The Effective Principal: A Research Summary.* Reston, VA: National Association of Secondary School Principals.

Seeley, C. (1993.) "Increasing Access or Ensuring Failure? Policy Makers Throw a Hammer into the Wall." *Algebra for the Twenty-first Century: Proceedings of the August 1992 Conference* (pp. 43-45). Reston, VA: National Council of Teachers of Mathematics.

Silver, E.A. (1995.) "Rethinking 'Algebra for All.' " *Educational Leadership*, 50(6), 30-33.

Steen, L. (1992.) "Does Everybody Need to Study Algebra?" *Basic Education,* 37(4), 9-13.

Improving K-14 Algebra Instruction: A Discussion of Teachers' Responsibilities and Students' Opportunities

Beatrice Moore-Harris
Houston Independent School District
Houston, Texas

We cannot have a discussion about how K-14 students learn without discussing how teachers teach. The improvement of students' learning of mathematics—algebra in particular—involves changes in student behavior as well as teacher behavior. To sustain changes in behavior, students and teachers must be able to reflect on new concepts, skills, and strategies as they practice them.

Let us first explore teacher behaviors. In order to develop teacher behaviors that facilitate students' ability to think algebraically, planning must be provided throughout the school year to debrief; to discuss successes and failures; to develop, review and revise lessons plans; and to continue professional development follow-up activities that address areas of student and teacher needs as identified by teachers. Professional development opportunities that address instructional alignment, appropriate use of data (including test-item analysis), and equitable as well as appropriate grading practices are non-negotiable components. Additionally, sessions on technology, both the calculator and computer, critical thinking, reading and mathematics, appropriate use of manipulatives, curriculum alignment, developmentally appropriate instructional practices, learning styles, and cultural diversity as they relate to and support algebra instruction should be included in all professional development activities.

Among the complexities and interactions of variables studied, many explanations and much confusion exist for the lack of student success in and preparation for algebra. There is clear agreement on one point, however; disparity begins to appear early in students' school careers (Camerlengo, 1993). As we begin to focus on solutions and strategies that address improving student achievement in algebra, it is important that we discuss how we may be able to provide a rich program in pre-algebra experiences that is taught in nontraditional ways. These pre-algebra experiences should increase the relevance of mathematics to the lives of the students by making it engaging and inclusive of cultural contexts that are born of student's environment, experiences, and interests. Furthermore, the use of activities that students find enjoyable, such as games, stories, dramatic play, and music, are critical instructional elements that incorporate what we have learned about learning styles and that bring mathematics to life.

Unfortunately, many students are trapped in a system that does not allow them to develop critical thinking abilities or to access a curriculum intended for preparation for algebra because the emphasis of instruction is on mastery of basic facts. The remainder of this paper will provide some brief suggestions about how we could attempt to redesign America's K-14 classrooms to give all students opportunities and experiences that allow them to change their behavior as learners and that specifically allow them to internalize concepts and skills that facilitate algebraic thinking. My intention is to provide a springboard for discussing current realities and the development of solutions to issues that are paramount to all who are in positions of responsibility—district curriculum supervisors, university-level instructors, public or private school building administrators, students, teachers, and parents.

Vygotsky (1978) developed the concept of the "Zone of Proximal Development" (ZPD) in which he made distinctions among the skills that a child demonstrates solidly on her own, those that she can demonstrate only with help, and those that she cannot demonstrate at all. It is the middle range (ZPD) that provides the most potential for growth. Often those who have just recently conquered a concept can best speak to the changes in understanding that have occurred. This middle zone can extend further up and further down than one may expect. On the upper end, some children can grapple with more complex ideas than they are generally given credit for. On the lower end, they often have not developed a solid understanding of even thoroughly explored concepts and, in new situations, need to remind themselves of what both we and they know they know. In schools with tracking, students in "lower" ability groups are not exposed to content that teachers perceive as too difficult for them. As a result, when students have developmental spurts, they may lose the opportunity to move ahead quickly because they are not being exposed to sufficiently challenging content.

Children in these groups also do not have access to participation in sophisticated discussion about topics: thus, they lose that positive modeling. The goal is not to have students reach closure on a particular concept; rather, the goal is for important questions to remain open so that students' understanding can evolve.

Using students' prior knowledge is a critical component of a successful mathematics program. The knowledge that students bring to a new experience will greatly shape their understanding of new concepts. Given a problem-solving task, students must bring to that task what they already know in order to develop solutions that are personally meaningful to them. Students may be asked to think of possibilities within a given situation and to create a word problem that involves real world contexts from their environment and experiences. This creates an opportunity for the students to explore language in a mathematical context by using language skills to discuss and create a world problem. By encouraging students to explain their thinking, we help them to solidify understanding. As they make deeper connections, develop new insights, and answer other students' questions, they further convince themselves of the truth of their understanding. Another technique is to remove the question from word problems. This forces the student to focus on the stated relationships in the problem, predict a conclusion, and then create a question that leads to a solution.

As mentioned earlier, games are of interest to most students and provide rich contexts in which students can explore mathematical ideas or practice mathematical skills (Bright, Harvey, and Wheeler, 1985). Many adults who are successful in mathematics had experiences as children where recreational time was spent engaged in games that provided practice in the acquisition of logical and analytical thinking abilities. Today's students do not come to school with these experiences because their recreational time is spent on activities that, in many cases, do not provide the same type of thinking. Thus, we need to provide recreational experiences in K-14 classrooms. Through the incorporation of additional experiences that are steeped in realistic and relevant contexts, the mathematics in a program becomes more accessible, and a more fertile foundation that promotes algebraic thinking is built. Concepts that are crucial to success in algebra, such as integers, fractions, the meaning of a variable, and ratio and proportion, are easily introduced, reinforced, reviewed, and maintained via a game format.

In addition, students need to learn to move from one representation to another using the language of mathematics. For example, students may use a manipulative to set up a problem, draw a picture to provide the teacher with evidence of what was done, use written or oral language to explain the process used and to summarize results, and, finally, use symbolic representation to state mathematically the problem and solution. To transition successfully from one representation to another depends on the concept involved and the students' experiences in mathematics, so the teacher must carefully address the matter in a way that provides adequate time for students of varied levels of understanding to make the transition.

As we look at the structure and organization of K-14 algebra instruction, we can consider the following example of how a typical class period is conducted during when the focus is to develop algebraic concepts, skills, and thinking processes. Imagine a classroom where students enter the room, are presented with a problem, and begin to work in small groups to solve it. The teacher monitors discussions, uses questioning strategies to assist students with their thinking, and organizes a discussion that allows students to share ideas and to summarize their findings or strategies. There is dialogue in which ideas are expressed in students' everyday language as well as appropriate mathematical language—written and verbal. Of course, there are times where direct instruction or review of concepts and skills will be appropriate and required. At all times, however, instructional decisions are made on the basis of what is best for meeting the needs of the students and careful analysis of student performance.

In conclusion, we must reflect once more on the National Council of Teachers of Mathematics' *Curriculum and Evaluation Standards for School Mathematics*. In particular, we should revisit the NCTM's five goals for all students:

- Learn to value mathematics
- Become confident in your ability to do math
- Become mathematical problem solvers
- Learn to communicate mathematically
- Learn to reason mathematically

As we continue the discussion of how to best provide K-14 algebra instruction for all students and how students best learn, these five goals are important to the discussion.

REFERENCES

Bright, G.W., Harvey, J.G., and Wheeler, M.M. (1985.) "Learning and Mathematical Games." *Journal for Research in Mathematics Education Research Monograph Series, No. 1.* Reston, VA: National Council of Teachers of Mathematics.

Camerlengo, V. (1993.) "Mathematics Specialist-Teacher Program: An Intervention Strategy for All (Reaching All Students with Mathematics)." Reston, VA: National Council of Teachers of Mathematics.

National Council of Teachers of Mathematics. (1989.) *Curriculum and Evaluation Standards for School Mathematics.* Reston, VA: Author.

Nelson, B.S., Silver, E.A., Smith, M.S. (1995.) "The QUASAR Project: Equity Concerns Meet Mathematics Education Reform in the Middle School." *New Directions for Equity in Mathematics Education.* New York: Cambridge University Press.

Vygotsky, L.S. (1978.) *Mind in Society: The Development of Higher Psychological Processes.* Edited by Michael Cole, Vera John-Steiner, Sylvia Scribner, and Ellen Souberman. Cambridge, MA: Harvard University Press.

*Synthesis of Day One**

Henry Pollak

Henry Pollak offered a summary of the issues raised on Day One of the symposium. He remarked, "One of the first things I noticed is a very healthy sense of rebellion among the group here. When you didn't think a question was the right question, you said so." He then provided a synopsis of responses to the following questions posed to the group on Day One.

Question 1: ***To develop coherence and depth in the algebra curriculum across K-14, what transitions do students need to make and over what period of time? Also, how long must students be engaged with an important mathematical idea so that understanding is achieved?***

Pollak noted that "the first transition is the one from the concrete to pictures of the concrete to symbolic representation. You go from play with things to a pattern to perhaps a numerical expression of that then to a symbolic expression of that, and all those together allow you to form a conjecture. With that conjecture, you can make predictions. Predictions give you control and that, finally, allows you to do verification or proof, depending on the level you are in. It's a very interesting sequence of events, and a number of you talked about it. You said, when you teach, don't try to go too fast. Students have to have a chance to prepare for these things. If, for example, you as a teacher want to use algebra tiles, students need to have played with things like that in the earlier years. You cannot suddenly hit students with algebra tiles, and say, 'See.' "

Pollak also noted the participants' interest in emphasizing links to the other parts of mathematics: "One of the important transitions, for example, is geometry and its connections to algebra. Explore an area, look at it in many different ways, persevere in what you and the students are doing, such as cooperative learning, use of English, and emphasis on reading."

In moving to the second part of **Question 1**, "How long must students be engaged with an important mathematical idea so that understanding is achieved? Pollak noted the participants' sense that this is the wrong question and that the straightforward response from many was simply, "As long as it takes."

Question 2: ***To develop criteria or lenses with which to select curriculum, appropriate pedagogy, and assessment, how should algebraic reasoning be defined? Also, how should state, national, or college placement exams be taken into account?***

* The following remarks by Henry Pollak and Gail Burrill were edited from verbatim transcriptions and serve as a transition from Day One to Day Two of the symposium.

Pollak said that "the answer I heard to the first part of Question 2 was that the transitions—more than anything else—help." He then discussed examinations—state exams, national exams, and college placement exams. He noted that "Everybody agreed that assessment drives instruction. They wish it were the other way around, but it isn't. As Hugh Burkhardt from England likes to say, 'What you test is what you get.' Something I hadn't realized is how important college placement exams are. People are more interested and more worried about these than the other exams. As someone here commented, if the college placement exam is deemed so important, what are we going to do about that at the K-12 level? Connections with the colleges are not good. The alignment between what goes on in the schools and what the college exams test is a real problem. There is a desire for teachers to be able to influence these tests in a serious way, including aligning the college placement tests with the *Standards*. It was noted that the SAT has changed, so why don't other tests change? At the same time that we need to know why each exam exists and what its goal is, others make the comment that students with real mathematical power should be successful with most tests, so quit worrying. Certainly this has been an important point of view in many of the curriculum revisions that I have seen in recent years."

Pollak also summarized reaction to the proposed national eighth-grade mathematics exam: "Most of you who talked about it were afraid of it. It was something that you were quite worried about. It may not go in the right direction. It may be at too low a level."

Question 3: ***To develop long-term support for teachers and* Standards*-based algebra, how much help do teachers need? Also, how do we develop public support for teachers and* Standards*-based algebra?***

Pollak said, "A major idea repeatedly presented was that teachers need time to talk to each other and to understand a district's scope and sequence. Elementary-, middle-, and secondary-level teachers in the system need time to work with each other and to get a holistic view of the curriculum. This is a very strong and useful idea. The general feeling was that high-school teachers need inservice primarily on pedagogy and elementary-school teachers need inservice primarily on content. Teachers for grades K to 5 particularly need materials that develop algebraic thinking. Coming from Teachers College as I do, I need to think about all that, too. We all need to integrate our efforts and to get support from families, administrators, and boards for what we are doing. But, specifically what do we do to develop support for teachers and standards-based algebra? Working with PTAs and business roundtables, writing newspaper columns, and gaining publicity for student work all were mentioned as ideas."

Pollak observed further: "One important thought about mathematics education is to involve employers. The system is too nearly a closed system. You need to get employers to think about the future rather than the past; what employees will need, not what they have needed in the past. You need to get them to think about how mathematics will be used rather than how they, themselves, learned it. That could be of great help. Please remember that standards and traditions in schools are not the same thing."

DISCUSSION RELATED TO JERE CONFREY'S PRESENTATION

In discussing Jere Confrey's presentation, Pollak observed: "Everybody agreed that there is no single approach to algebra that will serve the needs of all children. Participants expressed an interesting worry about the implications of saying that, and whether this means that we are going to have tracking. People wanted to say, 'No,' to that, but they were a little afraid to. What are the varieties of approaches? Cultures, expectations, assessments, families, abilities, pedagogics, lengths of class periods—these are all things that differ and are all things that affect how we learn and teach. What approaches should you use? Every approach you can think of was discussed today—the constructivist approach, the structural approach, and the application or context-driven approach, an approach through functions, an approach through connections to other parts of mathematics, an approach through group work, a traditional approach, a technology and math-lab driven approach, an approach through historical perspectives. These all have their place. The feeling was that you need a mixture of them."

Further, Pollak said, "Others noted that parents expect their children to have what they had. Remember the worry that if there should be more than one approach, that might automatically imply there is a need for tracking? Some people said, 'Hey, I don't like that word 'approach.' What does it mean? Is this a trap?' And they perhaps take care not to teach in a way that closes doors for any of their students."

"Another discussion question from Day One," Pollak noted, "was, What happens to the curriculum beyond the first course in algebra? The reaction was that this is the wrong question. If we are talking about the grades K through 14 curriculum, why is anyone even mentioning a first course in algebra as if it were gospel? Could we talk not about a first course but rather integration, as in the 'Interactive Mathematics Project'?"

Also, "another question was, What is the role and responsibility of teachers in changing success rates in algebra? And the reaction was to ask, what do you mean by success? Who defined it? Again, the answer seems to be, consider the transitions, consider discussions across grade levels, have the whole district work together, and so on. In staff development, people need to buy into the vision, and there needs to be involvement by teachers in the planning of the curriculum. School-university partnerships also are important for success, including those horrible placement exams. A major comment was that we all know what needs to be done. We should stop talking and do it."

AREAS NEEDING FURTHER DISCUSSION

"Last," Pollak observed that "the notion of algebra as a way of thinking versus algebra as an object of thought came up yesterday. I want to just ask you something. How much can you use algebra as a way of thinking before it has been an object of thought? How do you do that? What is the way in which the two of them can go together? I would like to hear some more discussion about that.

"I [also] want to mention that in three of the talks yesterday, piece-wise defined curves were mentioned. We had discussion of piece-wise defined curves on the CBL. We had discussion about renting videotapes and the different fees involved. We had discussion of the costs of the different ways of getting Internet services. Each of these is a piece-wise defined function. We do so little that is extremely practical in algebra—and in mathematics in general. The practical includes piece-wise defined functions. I think we ought to be paying a lot more attention to them."

COMMENTS ON HYMAN BASS'S PRESENTATION

Pollak closed his observations about Day One of the symposium by "making some comments about some of the things that were said yesterday, from Hyman Bass's talk in particular. Hy emphasized the many different interpretations of a mathematical operation. Take something like multiplication. You can interpret it as repeated addition. You can interpret it as counting arrays. You can interpret it as area. You can do all of these. The important thing is that some of those interpretations generalize, and some of them don't. Repeated addition is not easily generalized to fractions. Counting arrays is not easily generalized to fractions and real numbers. Area is. What you so often do in mathematics is have many contexts for an operation, and you look at them all. Then you generalize from the ones that generalize."

Pollak also noted that, "in the current rethinking of the *Standards,* one of the issues that people are working on is what is meant by algorithmic thinking. Hy gave a beautiful example of that in connection with the greatest common divisor. Yes, you can talk in terms of factoring the numbers and making a pool of all the factors that are in common. The trouble is that in the real world that is impractical. What is practical in the real world is, for example, the Euclidean algorithm. There are differences in the ability to carry out various algorithms. That is at the heart of algorithmic thinking. Hy also talked a lot about extending numbers in operations, so I want to add a thought to that. Say you start out with positive integers. You go from there to counting numbers. You go from there to all the integers. You go to the rationals. You go to the reals. You go to the complexes. If you include exponentiation, there are all sorts of stages in between. Each time you generalize, you increase the repertoire of operations, and you have to extend all the previous operations to the newly defined numbers."

Pollak added that, "the interesting fact is that every time you do this you lose something as well as gain something. It's like the economists say, 'There is no free lunch.' You push in here, and it comes out there. Every time you make an extension, you lose something as well as gain something. What do you lose when you go from the positive integers to all the integers? Well, the positive integers have the property that any non-empty set of positive integers has, i.e. a smallest member. Once you go to all the integers, that is not true. There may not be a smallest one in a subset of all the integers. What do you lose when you go from the integers to the rational numbers? When you take one integer divided by another, you no longer have a unique representation. Two over six and three over nine mean the same thing. Now you have lost uniqueness. As Hy pointed out, when you go to the reals, you have lost finite representation. You have, in fact, lost the ability to write down the number competely. As he said, if you then try to do multiplication by going from right to left like you are supposed to, the question arises of how you can start if there is no right end. So, you have lost something. When you go from the real numbers to the complex numbers, what do you lose? Order. There is no such thing as one complex number being less than or greater than another. This cannot be defined. In every one of the extensions that we talk about, there is a loss as well as a gain."

In a final comment, Pollak said that, in "addressing the differences between arithmetic and algebra—something we haven't talked about as much as I had expected—let me mention one aspect that has always bothered me. In much of arithmetic, all of the numbers are the same. There is no difference between 15 and 16 and 17 and 18. All the operations are carried out in the same way. In algebra, though, there is a big difference, because you have to look at things not only additively but multiplicatively. Interesting facts then are that 15 appears in the multiplication table once, 16 appears twice, and 17 doesn't appear at all. In arithmetic, the numbers are all the same, but in algebra, this knowledge of the biographies of the numbers is very valuable. When you start thinking about things like ratio and proportion, particularly in middle school, you are beginning to make the transition from thinking about just the additive structure into thinking about the multiplicative structure. That is part of the road from arithmetic to algebra."

Gail Burrill

OBSERVATIONS STIMULATED BY JOHN DOSSEY'S PRESENTATION

Gail Burrill noted that she had been thinking about what John Dossey said and about the notion "that we have lost a dynamic, growing view of what algebra is. I also am thinking about the observation that we are survivors and the notion that technology allows us to go back to a dynamic view. In my classes, using some of the technology has opened the students to thinking and developing and growing in their understanding of what algebra is all about, what solving equations is, how many different ways this can be done. So, it seems to me we can use technology as a way to get back to a dynamic, growing view. One topic that didn't get addressed in this technology conversation is momentum for change and the direction we take now. And how we have to be very careful that we are not just doing the old things in new ways. Why do I say that? Because I could already graph a function without technology. So, if I just keep graphing a function and graphing a function and just doing it faster using technology, I have not actually made the technology help me understand. I think we can use the technology to do old things in new ways, but we have to allow for the expansion of, for instance, the piece-wise functions that we were looking at earlier. That is something that was not actually in most of our curriculums—at least not until calculus—but it is now accessible to students. When they have access to these kinds of technology, students can think about and do things that were not part of the curriculum before."

TECHNOLOGY AND ALGEBRAIC REASONING QUESTIONS

Burrill said, "Let us look at algebraic reasoning and the question about what it is and how it differs from other lands of reasoning. I do not know if this was healthy rebellion or confusion, but when I read the algebraic reasoning papers, the coverage included patterns and generalizations and problem-solving and symbols and graphing and variables. There seemed to be a huge set of things that people think of as algebraic reasoning. In some of the discussion, I heard the questions, How do we know if we are seeing algebraic reasoning? Do we know it when we see it? There was a really intense conversation in one of the groups about just what does it look like when students are doing algebraic reasoning. If one is applying a definition, is that doing algebraic reasoning? I think we need to pay some attention to this."

Further, "When the question was asked, Does technology promote or inhibit the development of algebraic reasoning—and does the answer depend on the kind of technology employed—you all nearly unanimously said, 'Of course, sometimes technology inhibits, and sometimes it promotes. It depends entirely on the situation that you are working in and the goal that you have and how the whole thing is handled, what kind of training the teachers have, what kind of access.' "

In addition, Burrill noted that, "when the question was, How important is it to develop algebraic reasoning, everyone said, 'Well, that is a really silly question because of course it is very important to develop algebraic reasoning.' There was that healthy rebellion. When exploring the question of what reasoning and content skills should be in an algebra course or strand that promotes algebraic reasoning again, the responses were varied."

HIDING BEHIND EXCUSES

Burrill observed that, "one issue that came up in our discussions is that there is a tendency to hide behind excuses. I think that is what Bea Moore-Harris told us, and I think that is really important because it was reflected in the comments that I read. We can stack up a whole lot of excuses for why we cannot make things happen, then we can sit behind them and hope that someone else is going to solve them. We heard about the issue of articulation with the community colleges and the universities. Evidently this is a critical problem. We cannot hide behind it. We must do something about it. I am not exactly clear about what each of us can do, but I suspect that collectively in your states or in your communities you can start addressing this issue."

Burrill observed further that, "access to technology is an excuse that we can hide behind very easily, but there are states and there are cities that have made a difference in terms of giving every student access to technology. Every student that I know at my school has their own basketball—boys and girls. They can have their own calculator, too."

NATIONAL TEST

"Another question that came up involved the national voluntary test in mathematics," Burrill noted. "There were people who were very concerned that this assessment would set a minimum level, when, individually, each of us wants to be higher than the minimum. I would hope that, as a community, we can make sure that if we do have this kind of assessment—or if you have an assessment for your state—that it is not at a minimum level. There are two points here. One is what we may be after is a consistent vision, that there may be uniform kinds of things that students ought to be able to do at a certain point in time. The other point is that we need to be very careful about how we think about assessment: minimum level indicates that we are ranking kids, and I think that we do not want that to be the point of our assessment. I think we want more out of assessment than sorting students into piles. If we think about a minimum, we are not thinking about looking for what students know and are able to do. So, we need to push on that."

ADULT LEARNERS

Burrill observed that, "the following question came up once or twice, and it is really important: Do adult learners in community colleges or universities need a different kind of developmental structure to understand algebra? Or can you do the same things with them as with younger learners? This is critical because, as we learned yesterday, we have many students learning algebra when they come back to school as adults or even when they go right into college and have to take it over again."

PROFESSIONAL DEVELOPMENT

"The next-to-last topic I want to mention is professional development," Burrill said. "We need to have professional development, but we cannot have it for today's technology. I am sorry, given one of the comments I heard, but we should not teach every teacher how to use the CBL and how to use a TI85 or 86 because we do not know what is going to happen tomorrow. We cannot anchor the development of teachers to a particular piece of equipment at a particular point in time. This puts a really big burden on thinking about what is important in terms of content for teacher preparation because that has to transcend changing technologies. That is why this whole notion of symbols and technology is critical. We need to sort out what the technology is allowing us to do from what the important mathematics is going to be."

EQUITY

Finally, Burrill observed that, "equity issues were mentioned just once or twice, but it is important to recognize that teachers who tend to avoid technology are going to disenfranchise their students. This is a different kind of equity than we usually discuss. So, if there is a colleague who is not letting students use graphing calculators or computers, at some point, those students are going to suffer."

In addition, "another point I picked up from Bea Moore-Harris's talk is that it is really important for us to go beyond our own understanding. I had done the borders and blues problem that Betty Phillips shared with us. I had thought really hard about different ways to do that problem, yet I was amazed at all of the representations that came up. They never, ever would have occurred to me. I have to be able to move beyond my own understanding of how to fill in the holes so that I can let students have their choices of ways to think about things. When we have this notion of equity and the notion of trying to move forward, we need to put together technology, equity issues, and professional development in ways that will help us to help students understand algebra."

Presentations on Day Two

Capturing Patterns and Functions: Variables and Joint Variation (G. Lappan)

Functions and Relations: A Unifying Theme for School Algebra in Grades 9-12 (C. Hirsch)

Middle School Algebra from a Modeling Perspective (G. Kleiman)

Why Modeling Matters (L. Godbold)

Modeling: Changing the Mathematics Experience in Postsecondary Classrooms (R. Dance)

Algebraic Structure in the Mathematics of Elementary-School Children (C. Tierney)

Structure in School Algebra (Middle School) (M. van Reeuwijk)

The Role of Algebraic Structure in the Mathematics Curriculum of Grades 11-14 (G. Foley)

Language and Representation in Algebra: A View from the Middle (R. Billstein)

Teaching Algebra: Lessons Learned by a Curriculum Developer (D. Resek)

The Nature and Role of Algebra: Language and Representation (D. Hughes Hallett)

Capturing Patterns and Functions: Variables and Joint Variation

Glenda Lappan
Michigan State University
East Lansing, Michigan

In ordinary English context helps distinguish among possible meanings of common words. As a representation of ideas, a word stated free of some meaningful context does not communicate very well. For example, the definition of a word in the dictionary usually includes several possible meanings. The word used in context helps the listener or reader to differentiate among the various meanings and understand what the speaker or writer intends. In mathematics, we face the same dilemmas. Many mathematical words have different or at least different shades of meaning. In learning to understand both how to communicate in and how to decipher the language of mathematics, students have to determine meaning from contexts of use. Two of the key concepts in developing a deep understanding of functions are variables and joint variation. Variable is one of those mathematical words that has many meanings that must be determined from context of use.

In the new curricula that have been developed as a response to the NCTM's *Curriculum and Evaluation Standards for School Mathematics* and *Professional Teaching Standards for School Mathematics,* students learn mathematics through engagement with problems embedded in interesting contexts. This means that students have to learn to interpret different kinds of contexts. Students have to negotiate the "story" of the context—whether the "story" is from the real world, the world of whimsy, or the world of mathematics—to find ways to mathematicize the situation, presented in the story to manipulate the representations to find solutions, to interpret the solution in the original context, and to look for ways to generalize the solution to a whole class of problems. In addition, students have to interpret the context of the mathematization of the situation. Is the meaning of variable that of a place holder for an unknown? Or is the meaning of variable that of a domain of possible values for one of the changing phenomena in the context? Or is variable used in yet another way in the mathematization?

Joint variation of variables is the heart of understanding patterns and functions. As students grow in their ability to derive meaning for variables in contexts, they encounter variables that are changing in relation to each other. This pattern of change, or joint variation, becomes the object of study as certain kinds of change produce recognizable tables, graphs, and symbolic expressions. At the middle-school level, these important families of predictable joint variation are linear, quadratic, and exponential functions. In order to see how students can grow in their understanding of variable and joint variation, let us turn to a series of examples of significant stages of development in one of the reform middle-school curricula, the *Connected Mathematics Project.*

In the first situation given on the next page, students are challenged to determine variables of interest and to find ways to represent how these variables change in relation to each other over the day. The story line is that six college students are in the process of planning a five-day bicycle tour from Philadelphia to Williamsburg in the summer as a money-making business. They are exploring the route of the tour while gathering data on each day's trip. Day five encompasses a trip from Chincoteague Island to Norfolk, Virginia. The data are presented as word notes on the trip.

Malcolm and Sarah's notes:

- We started at 8:30 A.M. and rode into a strong wind until our midmorning break.
- About midmorning, the wind shifted to our backs.
- We stopped for lunch at a barbecue stand and rested for about an hour. By this time, we had traveled about halfway to Norfolk.
- Around 2:00 P.M., we stopped for a brief swim in the ocean.
- At around 3:30 P.M., we had reached the north end of the Chesapeake Bay Bridge and Tunnel. We stopped for a few minutes to watch the ships passing by. Since bikes are prohibited on the bridge, we put our bikes in the van and drove across the bridge.
- We took 7 1/2 hours to complete the day's 80-mile trip.

In this problem, the variables of time and distance traveled change in relation to each other but not in a mathematically predictable way. The elements, terrain, creature comforts, and rules of bridge use change the rates at which the distance is changing as time passes. The graph may be linear in parts, curved in parts, and constant in parts.

In the second situation, below, students are challenged to solve a problem that is stated in an open-ended way. To solve the problem, students have to find ways to represent and think about what the variables are and about two pairs of variables that are changing at different rates relative to each other.

Bonne challenged his older brother, Amel, to a walking race. Amel and Bonne had figured their walking rates. Bonne walks at 1 meter per second, and Amel walks at 2.5 meters per second. Amel gives his brother a 45 meter head start. Amel knows his brother would enjoy winning the race, but he does not want to make the race so short that it is obvious his brother will win. What would be a good distance for the race if Amel wants his brother to win but wants it to appear to be a close race?

Here students have to decide how to represent the progress of each brother so that they can determine what a good length would be for the race. Since the rate of change between time and distance for each brother is constant, each brother's time versus distance relationship is linear. The variables of time and distance in this situation change in a predictable way. One can predict the distance traveled for any number of seconds, which is a very different pattern of joint variation from the first situation.

Students can graph both functions on the same axes and see that the two lines cross. They have to figure out the significance of the point at which the lines cross and how this relates to determining a good distance for the race. Of course, students may reason from tables of data for each of the brothers. But, in any case, they have to identify the variables and make sense of which variable depends on the other and how the two pairs of variables relate to each other. Here the expression $2.5t$ can represent Amel's progress over t seconds. The variable stands for a whole domain of possible times. Students could write the function $f(x) = 2.5t$ or $d = 2.5t$ to show the independent variable, time, and the dependent variable, distance. They might also write $2.5t = 45 + 1t$ to show the equation that must be solved to find the point of intersection. Here the variable is standing for an unknown value, the value of t that makes the equation true.

In the third situation, below, students again have to deal with variables and the pattern of change between related variables, but the nature of the change or variation is different from either of the first two situations.

U.S. Malls Incorporated wants to build a new shopping center. The mall developer has bought all of the land on the proposed site except for one square lot that measures 125 meters on each side. The family that owns the land is reluctant to sell the lot. In exchange for the lot, the developer has offered to give the family a rectangular lot of land that is 100 meters longer on one side and 100 meters shorter on another side than the square lot. Is this a fair trade?

Here students may talk about the problem at a general level or solve the particular case by comparing areas. However, the question of whether the results are true in general for a beginning square of any size remains. Here the problem can be restated with smaller numbers as follows:

What happens if you own a square piece of land that is n meters by n meters and you are offered a piece of land that is 2 meters shorter on one side and 2 meters longer on the other? How does the area of the new lot compare to the original?

Here is a table that many groups of students make to record what is happening as the original square sides increase:

Original Square		New Rectangle			Change in Area
Length	Area	Length	Width	Area	
3 m	9 m^2	5 m	1 m	5 m^2	4 m^2
4 m	16 m^2	6 m	2 m	12 m^2	4 m^2
5 m	25 m^2	7 m	3 m	21 m^2	4 m^2
⋮	⋮	⋮	⋮	⋮	
n	n^2	$n+2$	$n-2$	$(n+2)(n-2)$	$n^2-(n+2)(n-2)=4$

The students notice that the change in areas from the square to the rectangle seems to be constant, 4 m. Since the students also develop a symbolic expression that tells what they do each time to find the change in area between the square and the rectangle, $n^2-(n+2)(n-2)$, the question of equivalence of expressions naturally arises. Why does $n^2-(n+2)(n-2)=4$? Students have various ways of looking at this equivalence. They can graph the data in different ways to observe the behavior of the graphs. There are different graphs that can be made from the data and that show different kinds of change. The functions, $A=n^2$, $A=(n+2)(n-2)$, where A is area and n is the length of the side of the square, show quadratic growth, and $D=n^2-(n+2)(n-2)$, where D is the difference between the areas, gives a constant value and hence a horizontal graph. But the students also are motivated to examine different ways to transform the symbolic expression. What is another way to express $(n+2)(n-2)$? By looking at a rectangle that is $n+2$ on one edge and $n-2$ on the other, students can use the distributive property to see that this expression is equivalent to n^2-4.

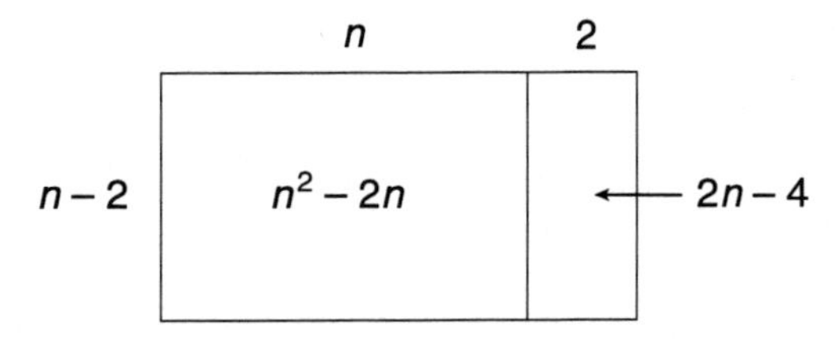

Therefore, the whole expression $n^2-(n^2-4)$ is always equal to 4.

The three problems I have presented here would be appropriate for different stages of a student's development of algebraic skill, but, nonetheless, all three serve to illustrate the centrality of variable and joint variation in understanding and using function to make sense of situations.

Functions and Relations:
A Unifying Theme for School Algebra in Grades 9-12

Christian Hirsch
Western Michigan University
Kalamazoo, Michigan

One of the most important transitions from middle- to high-school mathematics is the emergence of algebraic concepts and methods for studying general numerical patterns, quantitative variables and relationships among those variables, and important patterns of change in those relationships. The mathematical ideas that are central to that kind of quantitative reasoning are *variables*, *functions*, and (to a somewhat lesser extent) *relations* and their representations in numerical, graphic, symbolic, and verbal forms. Organizing school algebra around the study of the major families of elementary functions (linear, exponential, quadratic and polynomial, rational, and periodic) offers the opportunity to bring greater coherence to the study of algebra. Situating that study in explorations of contextual settings can provide more meaning to algebra and can provide a broader population greater access to algebraic thinking.

GOALS AND APPROACHES

From a functions and relations perspective, the continued study of algebra at the high-school level should enable all students to develop the ability to examine data or quantitative conditions; to choose appropriate algebraic models that fit patterns in the data or conditions; to write equations, inequalities, and other calculations to match important questions in the given situations; and to use a variety of strategies to answer the questions. Achievement of these goals would suggest that the study of algebra be rooted in the modeling of interesting data and phenomena in the physical, biological, and social sciences, in economics, and in students' daily lives. Through investigations of rich problem situations in which quantitative relations are modeled well by the type of function under study, students can develop important ideas of recognizing underlying mathematical features of problems in data patterns and expressing those relations in suitable algebraic forms.

Answering questions about the situations being modeled leads to questions such as the following, some of which are at the heart of a traditional algebra program. For a given function modeling rule $f(x)$, find

- $f(x)$ for $x = a$;
- x so that $f(x) = a$;
- x so that maximum or minimum values of $f(x)$ occur;
- the rate of change in f near $x = a$;
- the average value of f over the interval (a,b).

Early work by Fey and his colleagues at the University of Maryland (cf. Fey and Good, 1985; Fey, Heid, et al., 1995) using computer utilities demonstrated the promise of such a modeling and function-based approach. The emergence of hand-held graphics calculators puts such an approach in reach of all students and teachers of high-school mathematics.

Graphics calculator technology provides powerful new visual, numeric, and even symbolic approaches to answering questions such as those given on the previous page. The technology also facilitates exploration of more general properties of each family of functions and of all functions collectively; these can then be formally organized and verified at a later point in the curriculum.

A UNIFYING CONCEPT

Functions are a central and unifying concept of school algebra and, more generally, of school mathematics (c.f. Coxford et al., 1996). For example, symbolic expressions for function rules provide compact representations for patterns revealed by data analysis. Fundamental concepts of statistics, such as transformations of univariate or bivariate data, and of probability, such as probability distributions, are expressed and understood through the idea of function. The function concept can be generalized naturally to mappings such as $(x, y) \rightarrow (3x, 3y)$ or the following,

$$\begin{bmatrix} x \\ y \end{bmatrix} \rightarrow \begin{bmatrix} 0 & -1 \\ 1 & 0 \end{bmatrix} \begin{bmatrix} x \\ y \end{bmatrix}$$

which describe transformations of the plane. In discrete mathematics, early experiences with recursive descriptions of linear change (NEXT = NOW + b) and exponential change (NEXT = NOW × b) lead naturally to more general modeling with difference equations. Again, matrices are linked with transformations, and matrix methods are dependent on syntax and inference rules of algebraic symbolism. Finally, the mathematics of continuous change or calculus is fundamentally the study of the behavior of functions, including rates of change and accumulation.

SUMMARY

Organizing school algebra around functions and their use in mathematical modeling can provide a meaningful and broadly useful path to algebra for all students. Algebra as a language and means of representation is a natural by-product of this approach. Patterns that emerge through modeling with functions and studying families of functions can motivate at a later stage a study of the structure of algebra through deductive methods. Finally, the theme of functions and relations offers a way to provide a more unified approach to the high-school mathematics curriculum.

REFERENCES

Coxford, Arthur F., James T. Fey, Christian R. Hirsch, Harold L. Schoen, Gail Burrill, Eric W. Hart, Ann E. Watkins, Mary Jo Messenger, and Beth Ritsema. (1996.) *Contemporary Mathematics in Context: A Unified Approach.* Chicago: Everyday Learning Corporation.

Fey, James T., and Richard A. Good. (1985.) "Rethinking the Sequence and Priorities of High School Mathematics Curricula." In *The Secondary School Mathematics Curriculum*, 1985 Yearbook of the National Council of Teachers of Mathematics, edited by Christian R. Hirsch and Marilyn J. Zweng, pp. 43-52. Reston, VA: NCTM.

Fey, James T., and M. Kathleen Heid, with Richard A. Good, Charlene Sheets, Glendon W. Blume, and Rose Mary Zbiek. (1995.) *Concepts in Algebra: A Technological Approach.* Dedham, MA: Janson Publications.

Middle School Algebra from a Modeling Perspective

Glenn M. Kleiman
Education Development Center, Inc.
Newton, Massachusetts

To begin, we place algebra within the very general framework shown in Diagram 1 below.

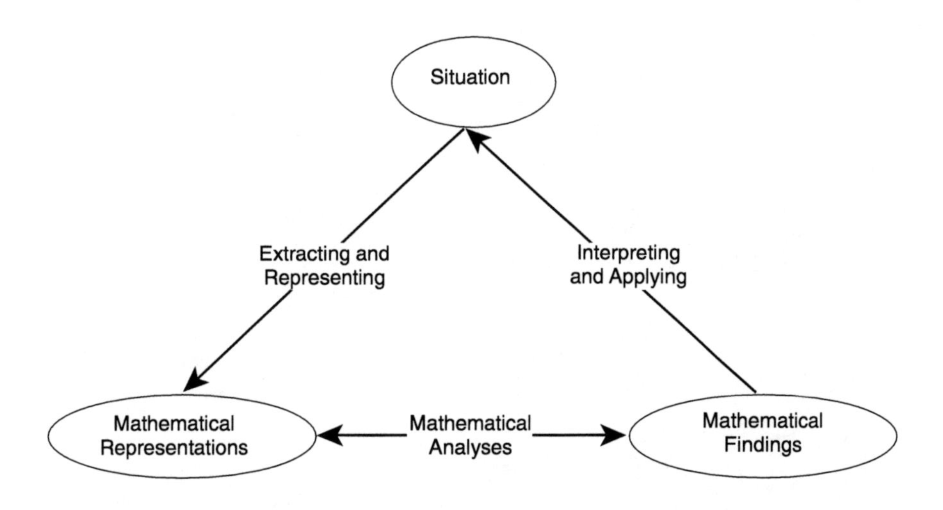

Diagram 1

This framework emphasizes that mathematics is more than working with mathematical symbols and tools. It also includes (a) extracting information from a situation and representing that information mathematically—the process of "mathematizing"; and (b) interpreting and applying mathematical findings to have meaning within specific situations.

This same framework can be applied to any area of mathematics. Algebra is defined by the representations, tools, techniques it provides, and the types of problem situations it enables one to address. Some specifics, focusing on algebra grades 6-8, are given in this paper. These are expansions of each of the three corners of Diagram 1 above.

The organizing theme of this framework is *modeling*. The other themes are incorporated within *modeling*. The *language and representation* theme is reflected in the processes of *extracting and representing*—bringing the original situation into a mathematical form—and *interpreting and applying*—translating back from a mathematical form to the situation. As we will see below in Diagram 3, the *functions and relations* and the *structure* themes are reflected in the *mathematical analyses* and *mathematical findings* components of the diagram.

First, let us expand upon possible types of situations in Diagram 2 below.

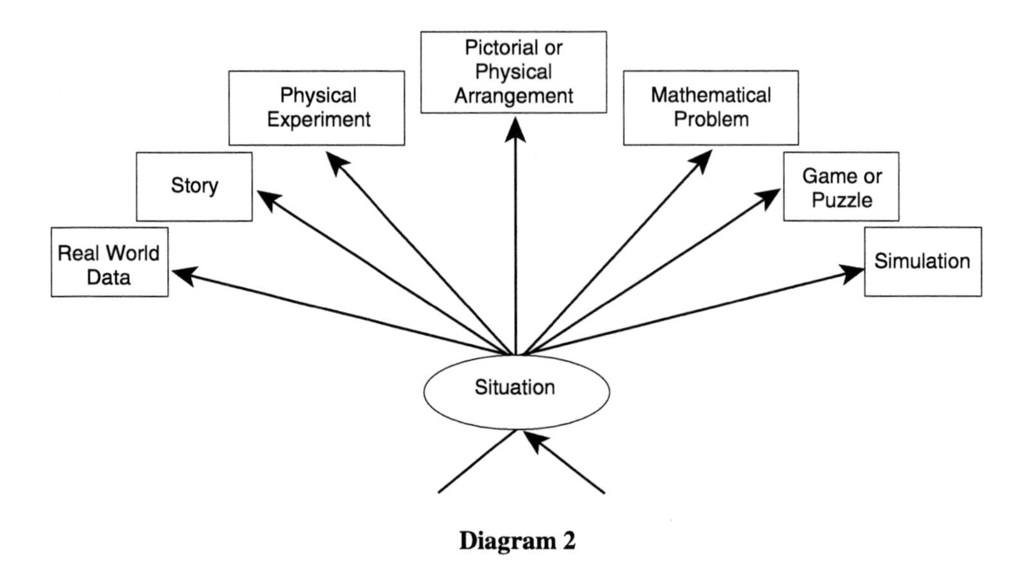

Diagram 2

The middle-school curriculum should include a wide variety of types of situations. An appropriate situation has the following characteristics: it is engaging for many middle school students; it can lead to significant mathematical explorations at an appropriate level of complexity; and it is manageable within the classroom.

Next, let's expand upon the *mathematical representations* for the *middle-school algebra* curriculum. The link across the representations in Diagram 3 below is a reminder that understanding the relationships among these representations is also important.

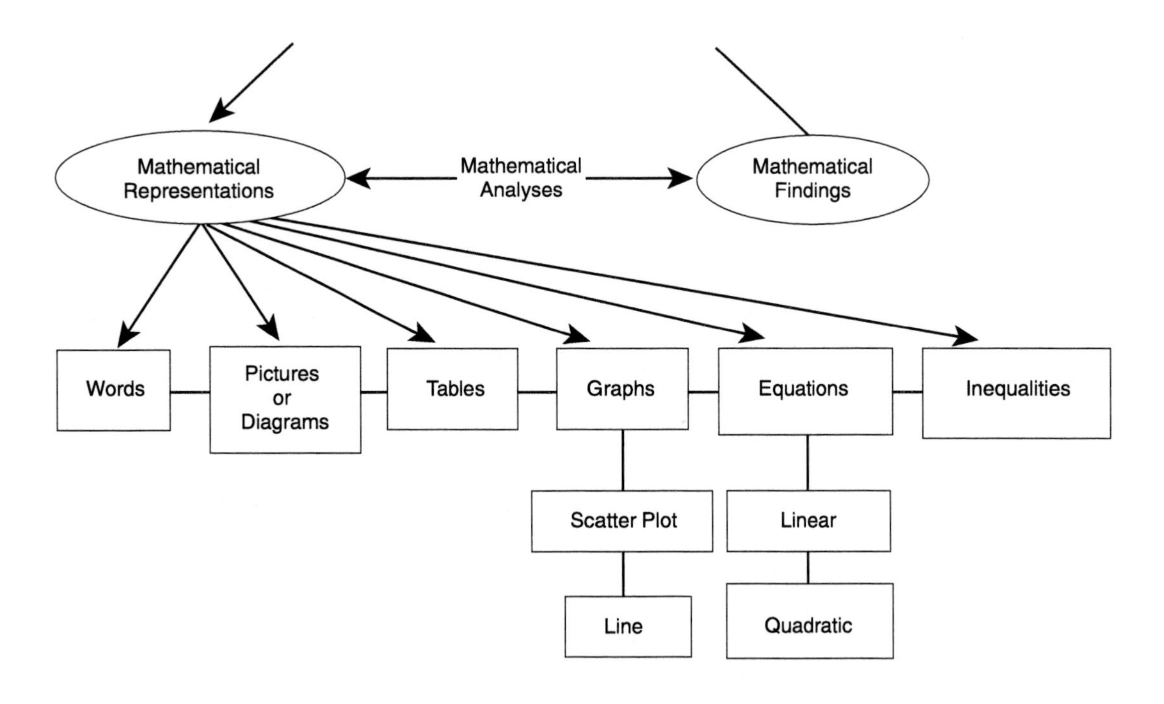

Diagram 3

Next, in Diagram 4 below are some of the types of *mathematical findings* we emphasize within the middle-school algebra strand

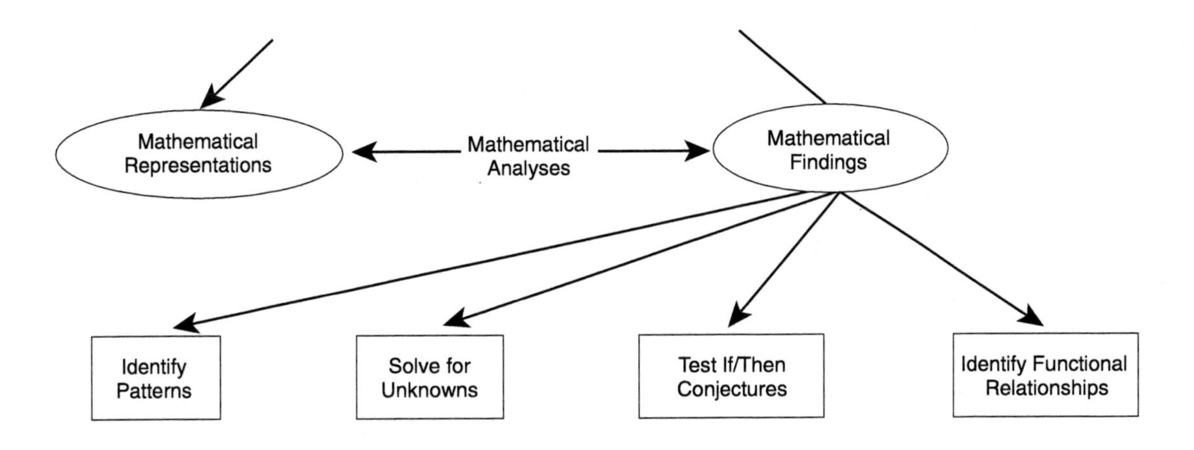

Diagram 4

Students should be able to use these findings to do such things as (1) use patterns to predict new cases in the situation; (2) interpret solutions of unknowns in terms of filling in missing information about the situation; (3) make if-then statements about the situation; and (4) use knowledge of functions to predict what will happen when one thing changes in the situation.

To get to these four types of findings, students need a repertoire of *mathematical tools and understandings*. Diagram 5 on the next page shows some categories of patterns students should understand; knowledge and techniques students will need to solve equations and inequalities; some tools for testing conjectures; and types of functions that students should become able to recognize and apply to understanding situations. All of these can be introduced in the middle-school curriculum, in many cases at an informal, context-based level, that forms a conceptual base for the more formal and abstract understandings that will develop in later grades.

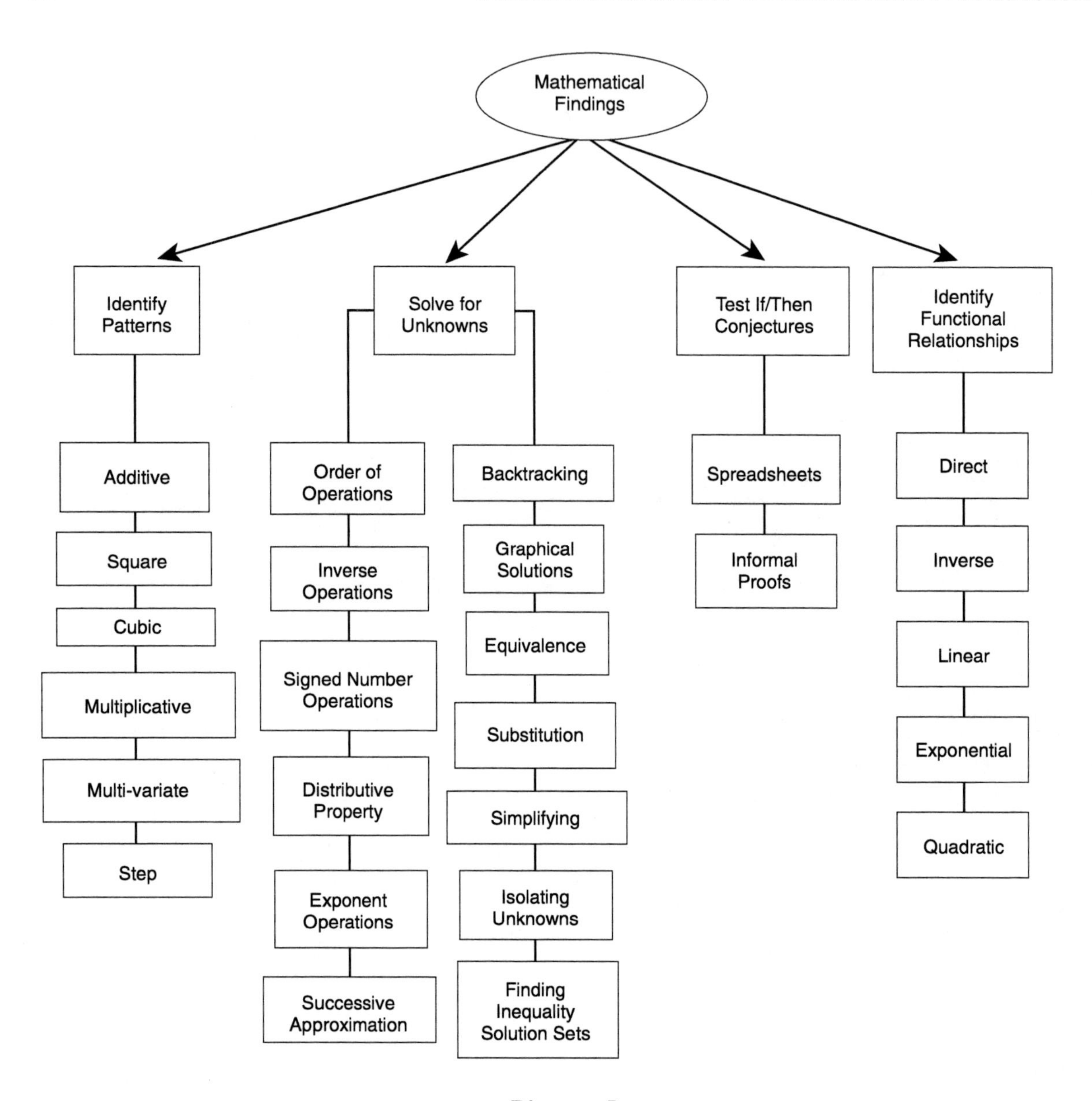
Mathematical Findings
Identify Patterns
Solve for Unknowns
Test If/Then Conjectures
Identify Functional Relationships
Additive
Square
Cubic
Multiplicative
Multi-variate
Step
Order of Operations
Inverse Operations
Signed Number Operations
Distributive Property
Exponent Operations
Successive Approximation
Backtracking
Graphical Solutions
Equivalence
Substitution
Simplifying
Isolating Unknowns
Finding Inequality Solution Sets
Spreadsheets
Informal Proofs
Direct
Inverse
Linear
Exponential
Quadratic

Diagram 5

Why Modeling Matters

Landy Godbold
The Westminster Schools
Atlanta, Georgia

For most people, mathematics is never used for its own sake, for its beauty. Rather, it serves as a tool for reasoning, for "getting answers." Thus, the applicability of mathematics is a primary reason for its existence, at least for many people. Of course, many people who reason quantitatively also would deny that they use (or can do) mathematics.

Application of mathematical ideas is not the same as modeling, but the two are related. And modeling is not just "curve fitting," although that may, at times, be one part of the modeling process. My guess is that there are at least as many descriptions of mathematical modeling as there are modelers. In general, however, there is agreement that modeling is a process and that it takes time; it is not something you just sit down and zip through. I would also argue that it is as much art as science; firm rules seem less appropriate than general guidelines. However, for the sake of having something to discuss, I will use the following description of the stages of the modeling process.

First, as suggested above, the following (or any) list of steps is not the "right" list. Second, if there is any "most important" guiding principle, it is likely "simple is better than complex." This principle has many corollaries that I will avoid trying to mention.

***Identify a situation*:** Notice something that you wish to understand, and pose a well-defined question indicating exactly what you wish to know.

***Simplify the situation*:** List the key features (and relationships among those features) that you wish to include for consideration. These are the assumptions on which your model will rest. Also note features and relationships you choose to ignore for now.

***Build the model*:** Interpret in mathematical terms the features and relations you have chosen. (Define variables, write equations, draw shapes, gather data, measure objects, calculate probabilities, etc.)

***Evaluate and revise the model*:** Go back to the original situation and see if results of your mathematical work make sense. If so, use the model until new information becomes available or assumptions change. If not, reconsider the assumptions you made and revise them to be more realistic.

"So where's the math?," I hear you cry. To varying degrees, it plays a role in steps two, three, and four, but it is central to step three. This step includes pretty much all of what most students would refer to as doing the math. Here, we take the key features and their assumed relationships and turn them into equivalent mathematical representations. Then those representations are re-expressed, while the integrity of the assumed relationships is maintained, until one or more representations tell us something about what we wanted to know.

To a large extent, this re-representation is a task ideally suited to what we call algebra. In fact, one view of algebra is that it is a language, with the additional property of permitting representations of ideas to be altered in ways that are known to preserve their meaning (even when they do not "look" the same). Mathematical tools to be used include probability, data analysis, functions, geometry, symbolic algebraic manipulation, graphs (of many types), matrices, and good old arithmetic, to name a few.

The process of moving from a set of assumptions to mathematical representations has long been a part of traditional school algebra through the dreaded "word problems." The modeling process extends this math-context interface in both directions. Up front, students must be able to examine a situation and make reasonable decisions about what is important and what is not. Thus, students must really understand something substantive about the context that is being modeled, as well as about the mathematical tools at their disposal. At the other end, in evaluating a model it is not good enough to realize that the results obtained do not match reality. To advance the model, it is necessary to know how the results fail to match, to what assumptions the differences are attributable, and what alternatives to those assumptions are reasonable.

Pedagogically, some of the implications of including modeling are obvious. Discussion of complex problems helps; group work is useful. Real situations are messy and involve numbers other than positive integers; calculators and computers are useful. Modeling solves problems that people care about; writing explanations of solutions to people is useful. Modeling requires serious consideration of the context itself; time spent on the "non-mathematical" aspects of understanding the situation and its constituents is useful.

Assessment, of course, cannot consist of a 45-minute, 50-item multiple-choice test. A list of short-answer questions is equally inappropriate. As noted above, modeling takes time and may involve collaboration. Assessment should reflect that. To quote Gene Woolsey, "The right answer unsold is the wrong answer." Modeling requires clear communication of the results and the reasoning behind those results. Assessment should reflect that also.

There are many payoffs for making modeling a major strand in the study of school mathematics. Students maintain and improve language skills. Students realize that there are serious uses for mathematics at all levels; you don't have to wait until you study calculus to answer meaningful questions. Students become better "math detectors"; they see opportunities for posing interesting questions in many non-mathematical settings. Students begin to appreciate the complexity of the world around them (physical, social, economic, etc.) and that "the right answer" depends heavily on the assumptions, stated and unstated, in "the question." Students learn persistence; serious modeling does not take place in a three-minutes-per-problem setting. Learning takes place across a broader time period; it is not possible to wait until the night before the test to begin trying to learn. Students develop connections, not just between math and contexts but within mathematics itself; modeling uses all the mathematics we know, and good questions beget new questions.

In addition to these payoffs, students learn to make decisions—about what to include as important features and assumptions, about what mathematics is appropriate to use, about the degree of agreement between reality and model—in an informed, defensible manner. This aspect is perhaps the most exciting for some teachers and students. This is modeling's answer to proof. Proof is just as vital as ever before, but now it may involve reasoning from perspectives other than only a set of axioms and is carried out using many forms of language (and never in a "T" chart). This need to construct valid arguments is frightening to many students, however, including some who have done very well in traditional mathematics. (True story: Several years ago, about November, I had a student ask in class, "Mr. Godbold, when are you going to quit trying to make us think and start teaching us math?") I have no data to back my opinion, but I believe that only in an educational system in which explanation is valued across the board, not just in isolated classrooms or by one or two teachers or just in geometry, can lasting progress on this front be made.

Perhaps most importantly, however, students realize that real mathematical inquiry is important, that they can do it, and that it is fun!

Modeling: Changing the Mathematics Experience in Postsecondary Classrooms

Rosalie A. Dance
Georgetown University
Washington, DC

Traditionally, the mathematics experience in college classrooms has been an empowering, mind-enhancing experience for only a small subset of students, and, traditionally, we have pretended to believe that the rest of our students lacked a certain *je ne sais quoi*. What we need to admit is that what we have been doing in mathematics classrooms has served to shut many people out. For example, for holistic learners, the intellectual process moves most effectively from a concern with the larger world and its problems to mathematics as methods of solution rather than from a concern with mathematical concepts and skills, first, and a faith, second, that there are problems that will submit to solution through application of these concepts and skills.

Appropriate models relate real problems to real-world issues. For example, application of exponential functions to population growth has been traditional, but now we also can use real population data to discuss current issues of overpopulation. We then can connect these issues to models of depletion of resources, such as petroleum (and, in many parts of the world, water) to allow study of significant problems. Research indicates that some of the demographic groups traditionally underrepresented in advanced mathematics courses and underserved by mathematics education are (speaking probabilistically) likely to be holistic learners.

Most models of real interest are more complex than traditional problems, and for effective investigation, they require students to work cooperatively. As students collaborate, share their insights, articulate their difficulties to each other and assist each other, they build a socially supportive setting for the learning of mathematics.

Mathematical modeling lends itself to hands-on experimental and investigative approaches with small groups of students working collaboratively. For example, we can simulate the dynamics of drugs being eliminated from the bloodstream with water and dye. The simulation clarifies understanding of the physical process and promotes ability to build the mathematical model. Hands-on models enhance understanding of foundation concepts for concrete learners. They provide a mental picture for repeated reference, not just for the investigation where they were used but for the class of problems about similar processes. Mathematical models of the dynamics of different drugs in the bloodstream and similar processes, such as elimination of pollution from lakes, can provide context for a range of algebraic concepts. With such models, students can answer not just the conventional questions (e.g., what is the equilibrium value for a certain drug dosage), but the real questions, the harder questions (e.g., what dosage is required to attain the necessary equilibrium value).

Traditionally, real-world context was limited by the "messiness" and complexity of real-world data and real-world processes. Now the use of technology as a tool for investigation and for calculation has freed us to work with real-world phenomena in a manner that does not always have to "ignore friction" and most of the rest of the environment. Technology enables us to explore mathematics in physical contexts in meaningful ways. A physical demonstration of light passing through water, some elementary geometry, and a graphing calculator provide the tools for a study of how changing parameters affect a mathematical model as well as give students the satisfaction of finding a reasonable estimate of the speed of light in water in terms of its speed in air.

Our goals for students ought to be somewhat loftier than for them to learn symbolic manipulation. The goals of a college mathematics department for students in their first two years might include (see also presentation overhead D on page 71).

- developing a habit of thinking mathematically (mathematical models of interest to students support this)
- learning to communicate mathematics
- developing an expectation of being able to make mathematical models so that they could answer questions and solve problems that arise in their own fields
- learning mathematical concepts in depth, with an understanding of their place in the logic of mathematics and their value in practice

If we do it well, modeling real-world problem situations has great value for the learning of mathematics. The models must be chosen carefully so that the mathematical concepts emerge as central. The teaching via modeling requires care; it requires built-in processes to assure that students have gained the mathematical knowledge inherent in the lesson and also the means to assess effectively what the student has gained. We can do this, and we must. If we do, more students will have a deeper understanding of algebra and its uses than has been the case traditionally.

MODELING: CHANGING THE MATHEMATICS EXPERIENCE IN POSTSECONDARY CLASSROOMS—PRESENTATION OVERHEADS

A. Mathematical modeling of real-world phenomena has the potential to effect profound change in the culture of mathematics classrooms.

Some aspects of traditional mathematics classroom culture:

1. Students work in isolation.
2. Learning through lecture.
3. Little incidental knowledge; little opportunity to think within the logic of the mathematical systems; few connections to other learning.
4. Thinking is often expected to occur within rigid, standard bounds.
5. Little context within which to assess usefulness or reasonableness.

B. Modeling can affect the culture of mathematics classrooms.

1. **Appropriate** models relate real problems with real-world issues.
   For holistic learners the intellectual process moves most effectively from a concern with the larger world and its problems to mathematics for methods of solution...
2. Most models of real interest are more complex than traditional problems; for effective investigation, they **require** that students work cooperatively.
   As students collaborate, share insights, articulate difficulties, assist each other, they build a socially supportive setting for the learning of mathematics.
3. Mathematical modeling lends itself to hands-on, experimental, investigative approaches.
   Enhances understanding of foundation concepts for both concrete and visual learners.
   Provides a mental picture for repeated reference.
   Enhances ability to develop abstract mathematical models.

C. Real-world context need no longer be as limited by the "messiness" and complexity of real-word data and real-world processes.

1. The use of technology as a tool for investigation and for calculation frees us to work with real-world phenomena without always having to "ignore friction."
2. Technology enables us to explore mathematics in physical contexts in meaningful ways.

D. The goals of a college mathematics department in the first two years might include the following:

- **Students develop a habit of thinking mathematically.**
  Mathematical models of interest to students support this.

- **Students learn to communicate mathematics.**
  Well-chosen models support collaboration.
  Investigation of real problems creates a need to write up work.
  A modeling environment helps students develop habits of reflection.

- **Students develop an expectation of being able to make mathematical models to answer questions and to solve problems that arise in their own fields.**
  Asking the "harder" questions supports this.

- **Students learn mathematical concepts in depth, understanding their place in the logic of mathematics and their value in practice.**
  Mathematical modeling provides an environment that enhances understanding of a mathematical concept and an immediate example of application.

E. Modeling real-world problem situations has great value for the learning of mathematics *if we do it well.*

- The models must be chosen carefully so that the mathematical concepts emerge as central.

- Teaching via modeling requires care; we need built-in processes to ensure that students have gained the mathematical knowledge inherent in the lesson and we need the means to assess it effectively.

We *can* do this, and we must. If we do, students will gain a deeper understanding of algebra and its uses than is traditional.

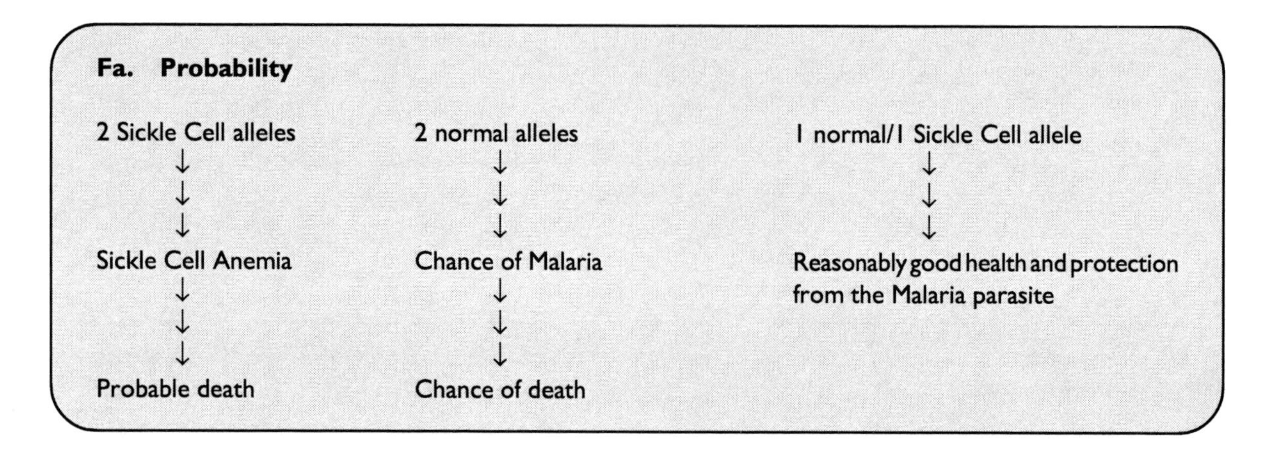

Fb. Meaningfulness and limitations of simulations and of predictions based on probability.

Mathematical model of fractions of alleles in adult population that are W given the fraction of the parent pool that was w.

Iteration of $w(n+1) = \dfrac{2-2(n)}{4-3w(n)}$.

Equilibrium—graphically, numerically, algebraically.

Solution of rational equation, $w = \dfrac{2-w}{4-3w}$.

Mathematical model for nb. of children in 1000 births who survive to adulthood.

Domain and range

Optimization

Quadratic function

Symmetry, zeros

Family of functions for varying malaria survival rates.

Algebraic Structure in the Mathematics of Elementary-School Children

Cornelia Tierney
TERC
Cambridge, Massachusetts

In looking at algebraic structure in elementary-school mathematics, I will focus here mostly on children's design and interpretation of graphs that show situations of growth or change. Examples will come from the work of children in classroom pilot studies of the "Investigations in Number, Data, and Space" curriculum (Susan Jo Russell, principal investigator) and in interviews on the research project, "Student's Conceptions of the Mathematics of Change" (Ricardo Nemirovsky, principal investigator).

Young children like to tell stories from their experiences and their imaginations. In these stories, children pay close attention to the order of events (and then..., and then..., and so then...) but seldom to the duration. Likewise, their graphs do not attend to duration of time. Rather than use a continuous scale to allow for a range of possibilities, they omit intervals that have no data in them. We call these graphs "data-driven." They tend to use natural language and additional elements, such as keys, to help the reader understand. In these graphs, children preserve all the data they think necessary to tell their story. (See Transparency 1.)

The structure of "system-driven" graphs is determined without concern for the particular data points. A systematic graph allows us to perceive patterns and generalizations in a universe of possibilities, not just those represented in a particular data set. However, in focusing on fewer variables and combining data into categories to fit to a system, information may be lost. For example, in showing changes over regular hourly time periods, one may omit the specific times that events happened within the hour. As long as the children's goal is to tell the story of the particular data, they see little value in a system in which data are lost. An issue for the teacher or curriculum developer becomes how to present graphing tasks that engage children and help them see the need to move toward more systematic representations.

One way to do this is to move attention from the specific data to the overall graph shape. In the "Investigations" curriculum, children are asked to make and interpret graph shapes that show growth of a plant, large population changes, or positions in a trip over time. No measures are written on the axes. The vertical dimension represents the height of a plant, the size of a population, the distance from the start; and the horizontal axis shows passage of time in days or hours. Such a task can be entered into at various levels of precision. When eight- and nine-year-old children first make or interpret a graph to fit a story of changes, they attend only to distinct heights or slopes on the graph without concern for the shape of the change in between. They identify places on the graph as meaning slow, medium, or fast. With experience, they learn to look at the shape of the change as well: is it "steady" or is it going "faster and faster" or "slower and slower"? (See Transparency 2.) By making communication with other students a main purpose, the need is created for more detail. Thus, the graphs children make, like the stories they write, need to be critiqued by others and go through several drafts.

Elizabeth's Graph of Changes in Population at Home

3:20 p.m. Elizabeth came home from school
3:35 p.m. Mommy came home 7
4:00 p.m. Elizabeth and Marie went out 5
4:30 p.m. Elizabeth and Marie came back 7
4:45 p.m. Cecily gets off the bus 8
5:30 p.m. Eben leaves 7
5:35 p.m. Cecily and Sam leave 5
5:45 p.m. Daddy comes home 6
5:55 p.m. We go to the airport 0
7:35 p.m. We come home from the airport 5
7:45 a.m. Daddy leaves for work 4
8:25 a.m. Eben comes 6
8:30 a.m. Eben's father leaves 5
8:35 a.m. Cecily and Sam come 8
8:37 a.m. Cecily's father leaves 7
9:15 a.m. Elizabeth leaves for school and ___ leaves 5
12:00 a.m. My brother comes home 7
12:30 p.m. Cecily and Sam leave 5

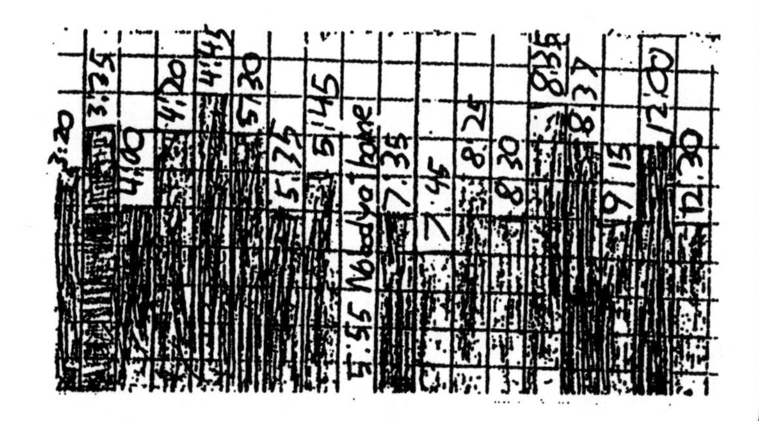

Transparency 1

Tell the story of the following graph of a plant growth.

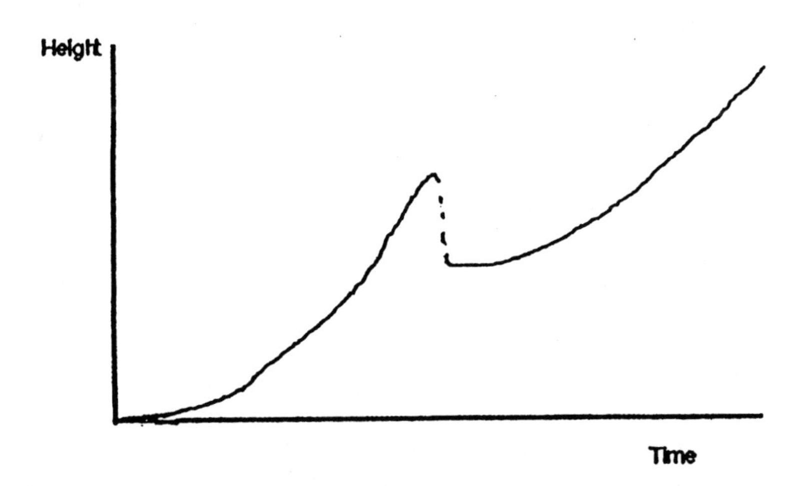

Responses from fourth-grade students:

- **Beginning it grows slow. Middle and at the end it grows fast.**
- **It went fast. Then it shrunk. Then it went fast and grew good.**
- **The plant started slow, then went fast, then all of a sudden it dropped (It most likely fell off), then it started up slow then it went very fast.**
- **This plant started out small, then it got bigger and bigger and bigger, and then it shrunk, but after a few days it started to grow again.**

Transparency 2 Plant Growth Story

We can pose questions so that students see the need to make more accurate drafts of graphs to communicate better, and we can suggest or model solutions. When third-graders show the motion of an elevator graphically with arrows on a vertical axis only, and others cannot tell where the elevator started and ended its trip, we tell them that mathematicians make graphs that show the order of events by moving from left to right. (See Transparencies 3a, 3b, and 3c.) When fourth-graders leave weekend days out of their graphs of plant heights, and we ask when the plant grew fastest, they realize that it appears that the plants always grew fastest between Friday and Monday. Children will also move toward more systematic representations in order to compare data or to combine data into one graph or table. Labeling an axis with the measured heights of one plant—1 cm., 1.5 cm., 2.5 cm., and 4 cm.—will not allow for a graph of a plant with different heights.

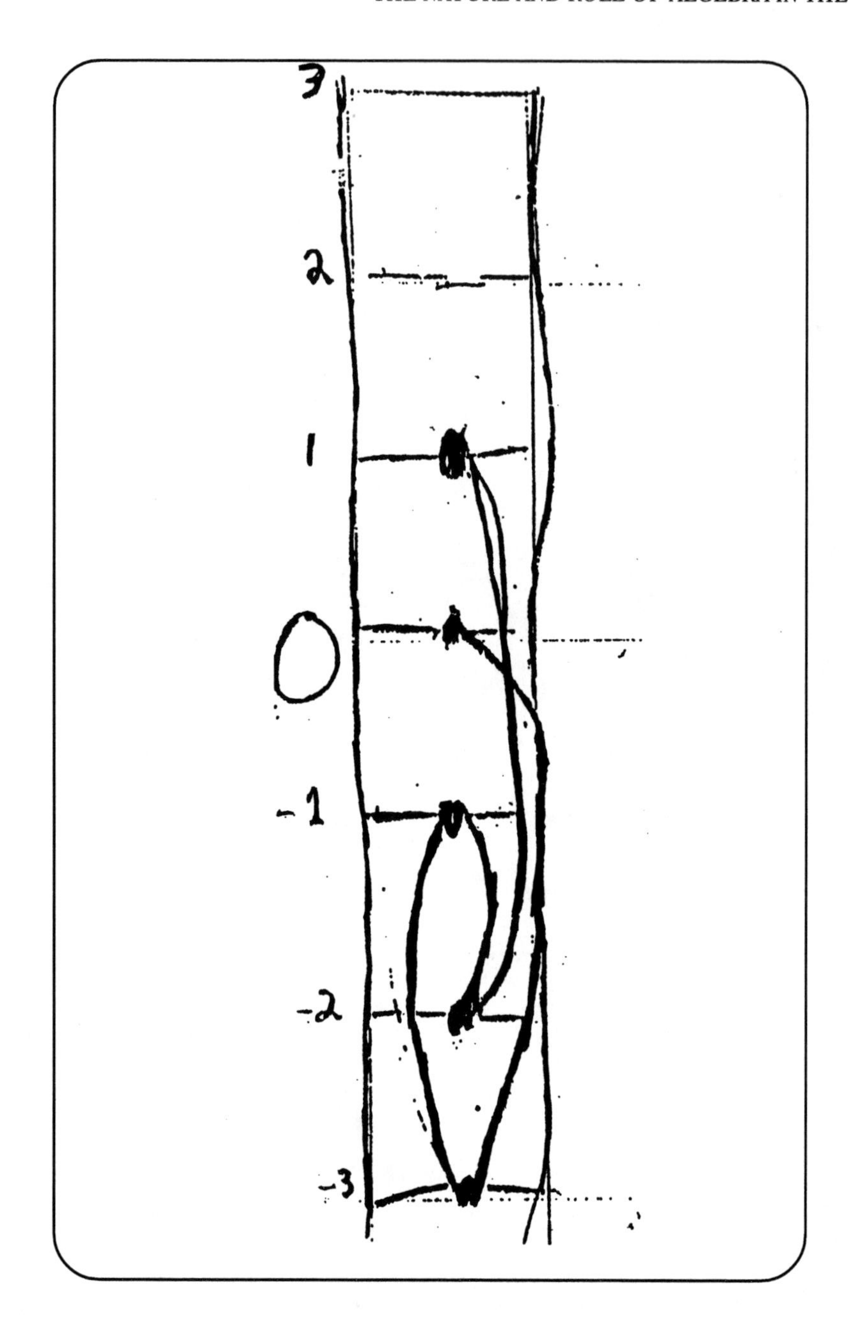

Transparency 3a Elevator Graph: Third Grader's Invented Graph

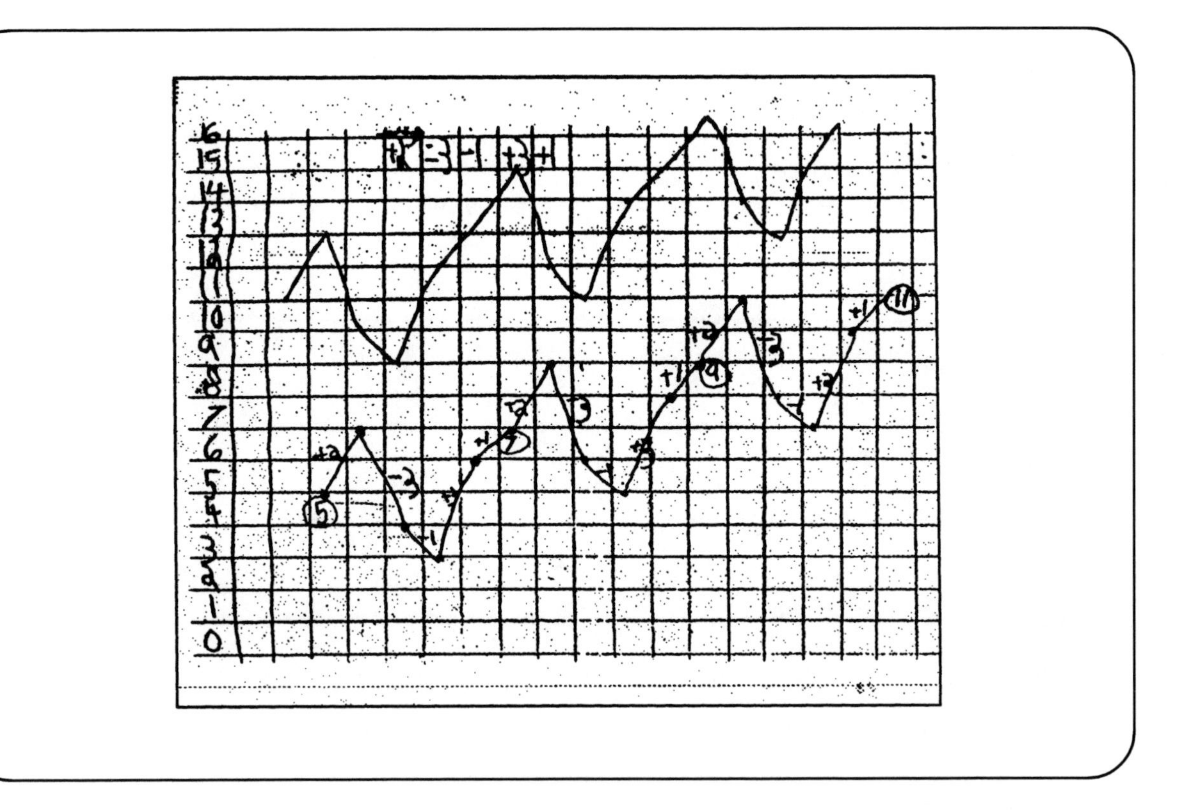

Transparency 3b Repeating Elevator Trip: Third Grader's Left to Right Graph

Net Change > 0
I am going to go higher and higher and higher and higher forever and ever.

Net Change < 0
I will go down and down and through the earth.

Net Change 0
I would keep going forward and forward and I would go out the side of the building.

Transparency 3c Descriptions of Elevator Trips with Repeating Sets of Changes

We also can embed conventions into the task, asking students to collect data at regular time intervals, to interpret a systematic graph, and/or to make a graph or table on a provided template. When templates are provided marked at regular intervals, most fifth-graders are able to invent data for tables or sketch graphs that fit with stories such as, "Run about halfway, then go slower and slower until the end." (See Transparencies 4a, 4b, and 4c.) We call these sorts of problems "construction problems" because students build their own solutions from among many, often infinite, possibilities.

Motion Stories

Story A: Run a few steps, stop abruptly, run a few steps, stop abruptly, walk to the end.

Story B: Run about halfway, then go slower and slower until the end.

Story C: Walk very slowly about a quarter of the distance, stop for 6 seconds, and then walk fast to the end.

Transparency 4a

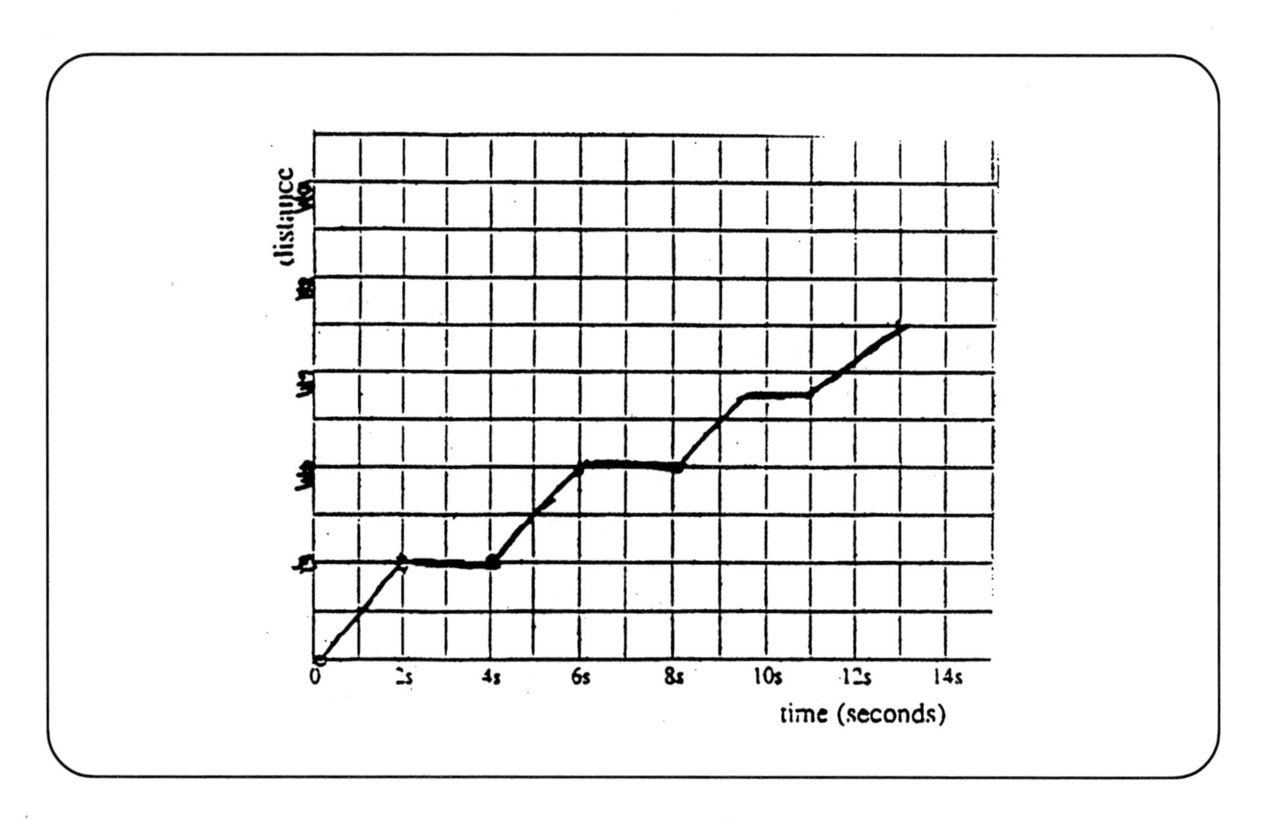

Transparency 4b Graph of Story B

Seconds	Distance
2	5 meters
4	8
6	10
8	11 1/2
10	12 1/2
12	13 1/4
14	14

Transparency 4c Table for Story A

As I look back at what I have written, I imagine that readers will wonder what this has to do with algebra. Where are the linear graphs for which equations can be easily found? We believe they should be introduced in early middle school, as one type of regular function among others, such as cycles or waves or graphs that go faster and faster or slower and slower. The significance of a pattern being linear or not depends on our expectations. In contrast to the many irregular possibilities that we come to expect, the regularity of a linear pattern we discover in a situational context is noticeable. For many algebra students, linear functions and their graphs are unremarkable, as they are the only ones worked with, and they appear on the textbook page devoid of context.

Experience of regularity begins in the early grades, when children create repeating patterns from objects and colors. Because these patterns offer a finite set of discrete alternatives, children are able to see and describe their regularity. At first they can only predict iteratively, but soon they make use of generalizations—every second or every third object is red, others are blue. (See Transparency 5.) They learn to see the similarity between two patterns made from quite different objects or from rhythmic sound and visual design. Notations themselves, such as ABAB or AABABCABCD, form patterns similar to the original design. Here again, as in the graphing, children must give up particular examples in order to describe and compare more generally.

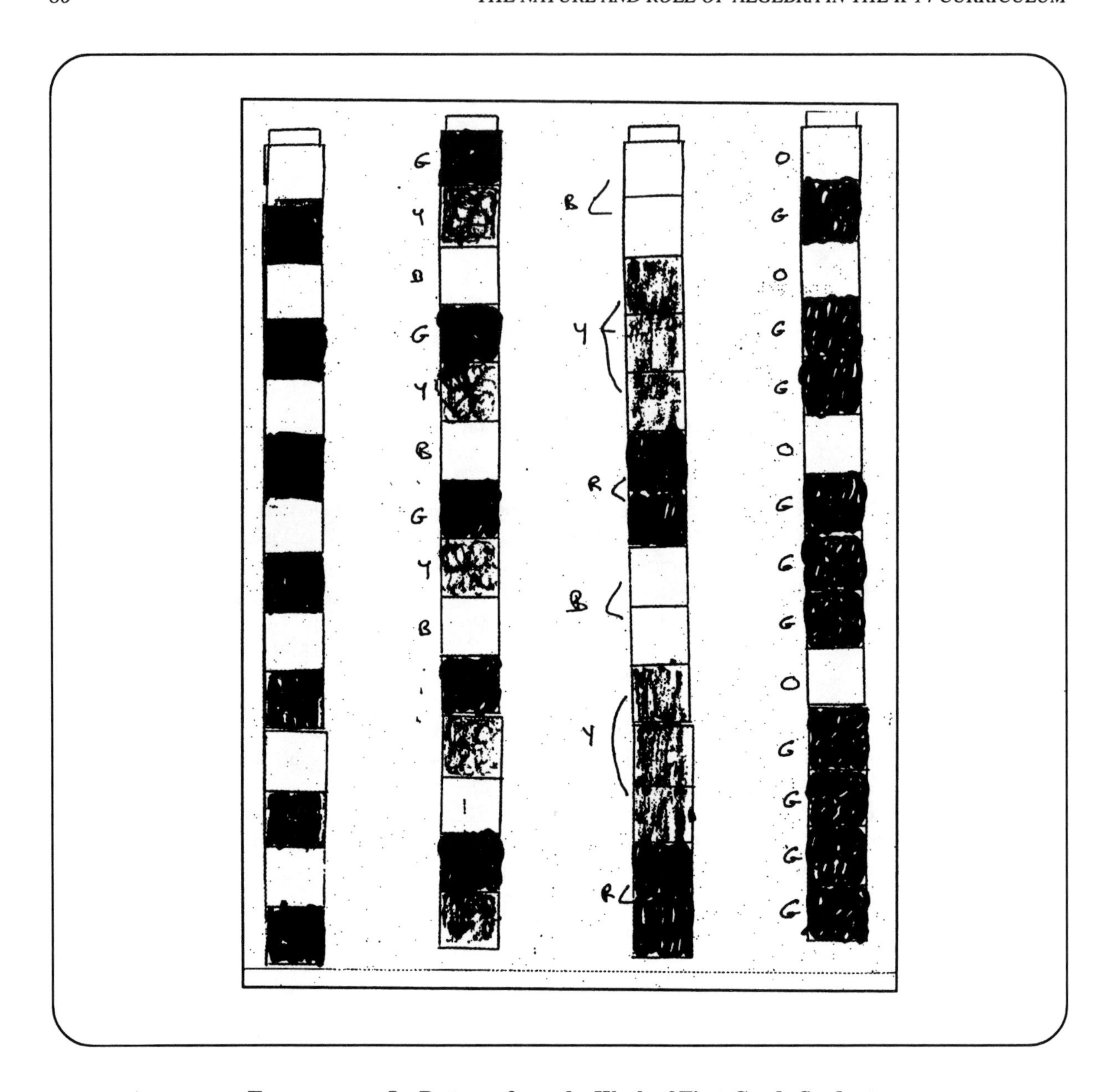

Transparency 5 Patterns from the Work of First-Grade Students

Finally, I will describe activities where children look closely at operations. One set of activities that parallels the graphing of changing situations is what we call "backwards problems," in which children undo additive changes. Strategies range from trial-and-error to undoing by adding opposites. (See Transparency 6.) Another activity where operations are examined is where children imagine they have a calculator with broken keys, and they create a number of related problems to get certain target numbers (answers) on the display. In the broken calculator work as with the graphing work, children are doing construction problems, finding for themselves patterns that allow them to make many problems of a similar type. (See Transparency 7.)

Backwards Problem

? + 1 − 2 = 3

Solve this missing information problem to find the place you started:

Place started	Changes	Place ended up on
	+1 −2	3

Responses from third-grade students:

- **I was going to do trial and era [sic] but the first move I made I got it right.**
- **I tried starting on three and I ended up on 2. So I thought if I started on three and ended on 2 I could start on four and end on three.**
- **You can do it backwards. If you went −2 and ended up on 3, you were at 5. If you went + 1 to get to 5 you started on 4.**
- **The net change is minus one. If you start on four and go minus one you get to three, so four is the answer.**

Transparency 6

Make 240 without using +	Make 16 using only subtraction
240 x 1 ÷ 1	84 − 68
240 x 2 ÷ 2	184 − 168
240 x 3 ÷ 3	1184 − 1168
240 x 4 ÷ 4	2284 − 2268
240 x 5 ÷ 5	9984 − 9968
240 x 6 ÷ 6	11184 − 11168
240 x 7 ÷ 7	99984 − 99968
240 x 8 ÷ 8	9999984 − 9999968
240 x 9 ÷ 9	
240 x 10 ÷ 10	
240 x 11 ÷ 11	
240 x 12 ÷ 12	

Transparency 7 Broken Calculator Problems

Structure in School Algebra (Middle School)

Martin van Reeuwijk
Freudenthal Institute
Utrecht University
Utrecht, The Netherlands

In the session on algebraic structure, I will focus on algebra at the middle grades and how algebra can be developed over time. Examples will come from the curriculum development project entitled "Mathematics in Context" (MiC). First, I will try to clarify the term "algebra."

ARITHMETIC, ALGEBRA, AND CALCULUS

In the discussions about "algebra for all" there is confusion about what is actually meant by "algebra." Algebra is the term that is used for school algebra; a domain within the mathematics curriculum as it is taught at schools (elementary, middle, secondary).

From a mathematical point of view, algebra deals with systems in which operations on objects play a role: addition (of numbers or other "things"), multiplication (numbers, other "things"), and the relations with the inverses. In other words, algebra deals with structure. Algebra is the study of operation structures. Following this point of view, arithmetic is a subdomain of algebra.

A common misunderstanding is that algebra is only algebra if insights in the structure are represented with letters. Also, it is thought that manipulating letters is the ultimate algebra. Another misunderstanding is that one can gain insight without understanding and without doing the mechanics.

Calculus deals with change of magnitude and continuous and discrete changes. Very large and very small are important in calculus; grasping the infinite small and infinite large is a way to draw conclusions about the finite space in between.

SOME EXAMPLES OF THE THREE DOMAINS

The following three examples are intended to illustrate the mathematical distinctions between the three domains of arithmetic, algebra, and calculus:

The fact that (1.41421)(1.41421) equals 1.999899241 is **arithmetic**.

An **algebraic** statement is that $(\sqrt{2})(\sqrt{2})$ equals 2.

A statement from **calculus** is that $\sqrt{2} = 1.4142135623730950488$.......

Graphs are part of **calculus**.

The fact that t^3/t_2 equals t is algebra, that t^3 for large t is much larger than t^2 is calculus, and that 1000/100 = 10 is arithmetic.

When talking about school algebra, I mean something other than mathematical algebra. Algebra in the context of school algebra is a coherent integration of elements from the three domains: arithmetic, algebra, and calculus.

ALGEBRA AT THE MIDDLE GRADES

Over the past five years, the Freudenthal Institute has been involved in a curriculum development project in which a complete, new mathematics curriculum named MiC has been developed for American students between 10 and 14 years-of-age. One of the content strands in this curriculum is "algebra," and it contains a collection of topics from different mathematical domains. In the algebraic structure session, I will use some examples to illustrate the philosophy and approach in this curriculum towards algebra. In this paper, I have restricted myself to outlining the philosophy and approach in general terms.

ALGEBRA IN MATHEMATICS IN CONTEXT

The algebra strand in MiC emphasizes the study of relationships between variables, the study of joint variation. Students learn how to describe these relationships with a variety of representations and how to connect these representations. The goal is not for students merely to learn the structure and symbols of algebra but for them to use algebra as a tool to solve problems that arise in the real world. For students to use algebra effectively, they must be able to make reasonable choices about what algebraic representation, if any, to use in solving a problem.

MiC APPROACH TO ALGEBRA

The MiC curriculum—especially the algebra strand—is characterized by progressive formalization. In other words, students rely heavily, first, on their intuitive understanding of a concept, then they work with the concept more abstractly. The realistic problem contexts support this progression from informal, intuitive understanding to a more formal, abstract understanding. Students can move back and forth from informal to formal depending on the concepts and the problem contexts. Their ability to understand and to use algebra formally develops gradually over the four-year curriculum. By the end of the four-year curriculum, students have developed an understanding of algebraic concepts and are able to work quite formally with algebraic symbols and expressions. Algebra in MiC lays a solid groundwork for mathematics at the high-school level.

EXAMPLE

Even and odd numbers can be visualized by dot patterns. Dot patterns also can be used to visualize and to investigate more complex (number) patterns. Symbols, expressions, and formulas (recursive and direct) can be used to describe the patterns. The formulas themselves can then become an object of study that lead to re-inventing such mathematical properties as distributivity and factorization. For example, when students investigate the structure of rectangular and triangular numbers, they can use visual representations to support finding appropriate algebraic expressions. In the algebraic structure session, I will illustrate this example with problems from the curriculum materials.

NO ALGEBRAIC STRUCTURES BUT STRUCTURE IN SCHOOL ALGEBRA

Algebra at the middle-grade levels builds on students' intuitive and informal knowledge of arithmetic, of symbols, patterns, regularities, processes, change, and so on—as developed in the early grades. Algebra in the middle grades does not need to lead to a complete and formal understanding of (the parts of) algebra. It is not the end of students' education. High school follows, and that is the place to formalize the concepts.

Algebraic structure—as described by Greg Foley, for example—is "number theory." We should be careful about making topics from number theory the focus of the mathematics curriculum, especially at the middle grades.

Factoring, divisibility rules, prime factorization, manipulating symbols and expressions, and other such topics have been the focus of the algebra curriculum, and students have not then had an opportunity to develop a meaningful understanding of the underlying concepts. The focus should be a long-term learning strand in which students can re-invent the algebra themselves, with the result being a mathematical system that is meaningful to students. The MiC algebra strand serves as an example of how this goal can be achieved.

The Role of Algebraic Structure in the Mathematics Curriculum of Grades 11-14

Gregory D. Foley
Sam Houston State University
Huntsville, Texas

MATHEMATICAL THEORY VERSUS RELEVANCE

The so-called "new math" movement of the 1960s brought such logic-based organizing themes as set theory and algebraic structure to the fore of school mathematics in the United States. Mathematical reasoning, axiomatic structure, and within-mathematics connections were driving forces of a reform movement motivated by American-Soviet competition and led by research mathematicians. The goal of preparing a cadre of highly capable engineers and scientists caused us to focus on the most able students. By contrast, the National Council of Teachers of Mathematics (NCTM) *Standards*-inspired school mathematics reform of the 1990s has been driven by calls for relevance—realistic applications, modeling, genuine data, and mathematics in context—and by a powerful collection of emerging instructional technologies. Mathematical communication, problem solving, and cross-disciplinary connections drive the current reform. The need for a generally well-educated population to remain competitive in a global economy has led us to conclude that "everybody counts" and that we need algebra for everyone. In the 1960s, we sought to motivate the mathematics; in the 1990s, we seek to motivate the students.

This, of course, is an oversimplification. A careful reader of the NCTM *Curriculum and Evaluation Standards* (1989) will notice an overarching theme of "Mathematics as Reasoning" and will see that the document says high-school students should learn about matrices, abstraction and symbolism, finite graphs, sequences, recurrence relations, algorithms, and mathematical systems and their structural characteristics, and that, in addition, college-intending students should gain facility with formal proof, algebraic transformations, operations on functions, linear programming, difference equations, the complex number system, elementary theorems of groups and fields, and the nature and purpose of axiomatic systems. The American Mathematical Association of Two-Year Colleges' (AMATYC) *Crossroads in Mathematics* (1995) contains similar calls for the content of introductory college mathematics. Ideally, there should be a balance between solid mathematics and relevance to the student and societal needs. The *Standards* documents for Grades 11-14 recognize this.

TECHNOLOGY AS CURRICULAR CATALYST

The influence of technology should not be downplayed. Technology is affecting the mathematics curriculum in several ways. Compared to the past, current technology gives students access to relatively advanced mathematical concepts and allows them to explore, describe, and display data with relative ease. Modern hand-held computers, such as the TI-92, have powerful features that allow students

- to operate with integers, rational numbers, real numbers, or complex numbers;
- to define, algebraically manipulate, graph, and tabulate functions of one variable, parametric equations, sequences, polar equations, and functions of two variables;
- to solve equations, find zeros of functions, and factor and expand expressions;
- to define, algebraically manipulate, graph, and tabulate sequences, polar equations, and real-valued functions of two variables;
- to operate on lists, vectors, and matrices whose entries are integers, rational numbers, real numbers, or complex numbers;
- to organize, display, process, and analyze data;
- to write, store, edit, and execute programs; and
- to construct and explore geometric objects dynamically and interactively.

In addition, modern technology and the related emergence of computer science make the knowledge of discrete mathematical structures more important. Technology has indirectly increased the use of statistics throughout society. It is no wonder, then, that the University of Chicago School Mathematics Project has a two-year sequence of *Functions, Statistics, and Trigonometry* (Rubenstein et al., 1992) followed by *Precalculus and Discrete Mathematics* (Peressini et al., 1992). There is simply more appropriate content after second-year algebra in preparation for postsecondary work in statistics, discrete mathematics, calculus, and linear algebra than in past decades. Technology makes this both possible and desirable.

WHAT IS THE ROLE OF ALGEBRAIC STRUCTURE?

Teachers of mathematics in Grades 11-14 must understand algebraic groups, rings, fields, and the associated theory. They need, for example, to recognize the importance of the complex numbers being a field and the significance of the fact that matrix multiplication is noncommutative and that matrices have zero divisors. They should see a logarithmic function as an isomorphism between groups and recognize geometric transformations as forming a group under composition. Furthermore, in keeping with the NCTM curriculum standards for college-intending students, high-school teachers need to be able to convey such understanding to their upper level students. This should be reinforced, amplified, and extended in lower division postsecondary mathematics courses, especially those in discrete mathematical structures and linear algebra.

We must, however, be careful not to make algebraic structure the overriding focus of mathematics in Grades 11-13, except, possibly, for the most gifted and talented students. On the other hand, it is essential that, in Grades 14-16, students acquire a clear vision of the "big picture" provided by a structural understanding of algebra. While we help students acquire this vision, we continually should call their attention to the numerous specific examples of groups, rings, and so on, as they learn the common structure and associated theory. There are abstract algebra textbooks, such as Fraleigh's (1989), that do a good job of this.

Abstract algebraic structures can serve as important organizing tools for the mathematics curriculum, but we should not fall into the trap of creating a new "new math."

REFERENCES

American Mathematical Association of Two-Year Colleges. (1995.) *Crossroads in Mathematics: Standards for Introductory College Mathematics Before Calculus.* Memphis, TN: Author.

Fraleigh, J. B. (1989.) *A First Course in Abstract Algebra, 4th Ed.* Reading, MA: Addison-Wesley.

National Council of Teachers of Mathematics. (1989.) *Curriculum and Evaluation Standards for School Mathematics.* Reston, VA: Author.

Peressini, A. L., et al. (1992.) *Precalculus and Discrete Mathematics* (University of Chicago School Mathematics Project). Glenview, IL: Scott-Foresman.

Rubenstein, R. N., et al. (1992.) *Functions, Statistics, and Trigonometry* (University of Chicago School Mathematics Project). Glenview, IL: Scott-Foresman.

Language and Representation in Algebra: A View from the Middle

Rick Billstein
Director, STEM Project
University of Montana
Missoula, Montana

Algebra can be thought of as a language, and students learn language best in their early years. We should begin teaching informal algebraic concepts in the elementary grades and continue to develop the concepts throughout the middle-school years. Algebra too often is taught as rules and tricks without an understanding of the concepts. Then the jump to the formal level is often made too quickly for the concepts to be mastered. Topics should be developed slowly and informally without symbol manipulation as the primary goal. The jump into symbol manipulation should come only after students recognize the need for it.

Algebra has been described as a way of thinking about and representing many situations. Unfortunately, many textbooks confine algebra to solving equations and manipulating symbols. Other representations, such as graphs, tables, patterns, diagrams, and other visual displays, should be used as appropriate. Visual representations are powerful because they help abstract mathematical ideas to become concrete. Since different representations may provide new or fresh insights about a problem, each representation is important and plays a role in the learning of algebra. There should be many opportunities for students to make transitions between the various representations. As students mature mathematically, they learn which types of representations are most useful for which kinds of problems. Students need to describe various representations in their own words. After a representation has been used, it is important to discuss it in terms of the original context. Interaction between teachers and students is important in development of language and representation skills.

At the middle-school level, students might be asked to translate between words, tables, graphs, and equations. Given any one of these representations, they could be asked to determine any of the other three. Having students work in groups and share representations makes them aware that different representations can be equivalent yet look quite different. This is a powerful experience in middle school and will pay huge benefits at the high-school level. Instead of always translating from words to representations, as in the traditional curriculum, we now ask students to translate representations to words.

With the use of technology on the rise, new understandings of symbol manipulation are needed to model situations that can be entered into a computer. For example, spreadsheets can be used to analyze complex numerical data from a problem situation. Algebra becomes important because it is the language used to communicate with the technology. Spreadsheet formulas are but one example of a form of algebra. Technology allows students to experiment, to investigate patterns, and to make and test conjectures. Technology allows us to go where we could not go before because the mathematics became too "messy." Students need experiences making representations with and without technology.

Students should be involved in "doing" mathematics at the middle-school level. It is important that they investigate problems and be involved in hands-on activities. "Doing" mathematics provides students with opportunities to communicate about algebra. There is little communication in a typical algebra textbook. The language that students use will develop as they become more mathematically mature. Curriculum materials not only must contain good problems, but also the problems must be structured in such a way that solving them will help students achieve the desired learning outcomes.

New materials are now becoming available in which algebraic ideas are taught throughout the curriculum. Teachers must be made aware of these materials and support must be provided to train teachers to use these materials. Teachers must understand that when we say we want to include algebra in the middle school, we are not talking about the algebra that they had when they were in school.

Hugh Burkhardt of the Shell Center for Mathematics Education in England describes algebra as "inherently slippery" and has said that "having separate algebra courses is one of the United States' great self-inflicted wounds" (1997). Most National Science Foundation (NSF)-funded middle-school curriculum developers have struggled with this and are constantly asked about the role of algebra in the materials and how their curricula fit with an Algebra I course. The "Six Through Eight Mathematics" (STEM) project response has always been that algebra should not be a separate course taught at a particular grade level but, rather, that it should be a strand taught within the mathematics curriculum at every grade level. The teaching of algebra should be integrated with the teaching of other mathematical strands, such as statistics or geometry.

The traditional Algebra I course should not be a required eighth-grade course as it is in many schools because this means the sixth, seventh, and eighth-grade curriculum must be covered in only two years. Two years is not enough time to develop adequately topics in probability, statistics, measurement, discrete math, number theory, and geometry. Many of the negative feelings that develop towards mathematics as a result of an "algebra course" might be eliminated if the algebra were integrated into the curriculum as a strand. Students would no longer remember algebra as a course in manipulating expressions and solving symbolic equations. Student experiences in middle-school mathematics courses might then actually prepare them for—and encourage them to take—additional mathematics courses, especially if those courses were taught in the reform mode of the new NSF high-school projects.

STEM has found that one way to teach algebra effectively is to make it useful to students. Real contexts that are meaningful to students play a major role in algebraic learning. Real contexts do not mean that all problems have to come from students' everyday lives but, rather, that problems must make sense to students. Algebraic abstraction is motivated by the need to represent the patterns found in the context. Algebraic thinking is more important than algebraic manipulating. To develop algebraic thinking, we need to include informal work with algebraic concepts in the middle school and not move too quickly to the abstract level. For example, being able to set up graphs or tables in various problem settings brings mathematical power and understanding to students. Experiences with graphs should include a detailed plotting of points to determine a graph as well as experiences with the overall shapes of graphs based on the information in the problem. If students are given a graph, they should be able to write a story about it. Having students communicate about mathematics is a worthwhile goal in the new middle-school curricula.

Don Chambers wrote, "Algebra for all is the right goal at the right time. We just need to find the right algebra" (1994). The NSF-funded middle school projects are taking us closer to finding the right algebra.

REFERENCES

Burkhardt, H. (1997.) Personal Conversation at the National Council of Teachers of Mathematics National Meeting in Minneapolis, MN.

Chambers, D. (1994.) "The Right Algebra for All." *Educational Leadership*, 51, 85-86.

Teaching Algebra: Lessons Learned by a Curriculum Developer

Diane Resek
San Francisco State University
San Francisco, California

In this symposium, I am approaching the issue of teaching algebra from the perspective of a curriculum developer. I am one of the developers of the National Science Foundation's (NSF's) secondary curriculum, "Interactive Mathematics Program" (IMP). The original design for the program was based on the experiences of the developers in past curriculum projects and in their own teaching. In this new project, we have had experience teaching algebra to students in elementary school, high school, and college. The original IMP curriculum has been rewritten three times—shaped by the authors' observations of the curriculum as it was taught in different classrooms, by the comments and suggestions of teachers, and by student work.

What follows are some statements about what I now believe about the teaching of algebra.

INTUITION SHOULD COME FIRST

People often ask for evidence that the new curriculum projects "work." There is some evidence of this, but there are mountains and mountains of evidence that the traditional methods don't "work." Now, exactly why the traditional curriculum does not work is open to question. My personal belief is that the chief culprit is the teaching of manipulative skills in a way that does not allow adequate intuition to come into play about what the symbols mean and why the manipulation is valid and useful.

It is not that we do not know how to teach manipulation in meaningful ways. Many projects have shown us how to do this for years. One way to develop intuition about manipulating equations symbolically is by tapping into both the students' familiarity with the fact that equations are statements about functions and the students' comfort in associating the symbolic form of functions with other representations of functions, graphical or numerical. Student familiarity and comfort must be developed over time.

It is my hope that much of the work in elementary-school and middle-school algebra will be on developing student comfort in moving between representations of functions. In general, we can decide what we need to teach at various levels by looking at what is difficult to teach later on. Traditionally, we have looked at what skills we thought were needed for success at one level and then taught those skills at the lower level. I am suggesting that we look at what skills or understandings are difficult for students at one level and try to develop intuition at lower levels that might serve as a basis for those skills or understandings.

UNDERSTANDING DOES NOT COME IN DISCRETE PACKAGES

One difficulty with building students' use of intuition is that this requires time—often several years. Teaching one concept over time conflicts with the traditional idea of organizing teaching around subject matter. Traditionally, one studies a chapter on linear equations at one time and that subject is then checked off. However, if we want students to understand linear functions in different representations, this must be worked on over several years. Striving for long-term exposure to subject matter creates a difficult bookkeeping problem. It becomes difficult to check off the skills that students have. As the public pushes us for accountability, they push for a neat and tidy assessment system. Unfortunately, that kind of assessment system does not match the way students learn.

At any level—first grade, high school, or college—students do not study a deep topic and suddenly "get it." Understanding comes gradually. It develops over time. Anyone who has asked students to write about a topic knows this. Reading what they have written, we can see that there are things the student seems to understand and other things he or she has not yet come to terms with. Rarely do we get a picture of perfect understanding. Our curriculum must be structured so that students can work on tasks at different levels and so that everyone in a class will grow.

I am not saying that teachers should not be accountable for students' learning or that students should not eventually master key ideas. I am saying that we need to take into account how students learn and that this is gradual. We must not let the difficulty of assessing understanding sway us from trying to teach effectively.

WE CAN LEAD STUDENTS TO WATER, BUT ...

Once we have decided what we want to teach and how we want to teach it, we have to wrap it in the right paper. This is not because students are lazy or do not have good taste. It is because their minds cannot actively work with material if they have no way to relate to it. Contexts and relationships to other subject matter can provide students with a door to approach new mathematics.

This is not to say we should never teach mathematics without a "real life" context. I am saying that we have to introduce the mathematics in a "real life" situation that students can relate to or in the context of other mathematics that they are working with. Once they have gone into the mathematical ideas, they can and will go on to work on the "bare" mathematics.

A few students do think well symbolically and do not require much of a context. Most of us here were that kind of student. In the past, success in algebra was reserved for us and others like us. Algebraic knowledge is too important to reserve for so few. It also is not clear that people who think in this way have the most to contribute even to pure mathematics. We have to open the doors to others. It is not that hard to do.

The Nature and Role of Algebra: Language and Representation

Deborah Hughes Hallet
Harvard University
Cambridge, Massachusetts

Students arriving at college should be familiar with verbal, symbolic, graphical, and numerical representations. They need to be able to manipulate each one and be able to convert one to another. Manipulating each of the representations requires some degree of technical skill supported by conceptual understanding. This understanding must comprise both an understanding of how the representation works in general and of each particular object being represented.

To work with graphs, students need to understand how graphs are constructed. For example, they need to understand that values of inputs to a function are measured horizontally, whereas values of outputs from the function are measured vertically. This will enable students to interpret intercepts and to estimate values of limits and asymptotes. Technology is changing how much technical skill students need to have in drawing graphs by hand, but it has not changed the fact that students need to understand how the zeros and symmetries of a function appear on a graph, where to expect asymptotes, and what sort of scale will show all the features of the graph.

Numerical data, usually in a table, is for many students the least familiar representation. Students need to understand how the data were generated (from an experiment, by using a formula, for example). They should be able to work with numerical data, such as rounding, interpolating, and extrapolating (where these make sense). The ability to find patterns in data, such as where values are increasing or where there are constant differences, is a useful skill.

The manipulation of symbols traditionally has formed the largest part of an algebra course. It is still central. Students must be able to solve equations, collect terms, simplify, and factor. The degree of skill and the speed required may be altered by changes in technology. For example, methods of factoring higher degree polynomials are probably not as important as they used to be. However, what it means to factor a polynomial (for example, that it is not usually useful to write $x^2 - 2x = x(x) - 2x$) is as important as ever. Experience and observation will suggest the most effective balance between paper-and-pencil work and technology. Currently, there is a wide range of opinion about the best way to develop manipulative skill, ranging from not allowing technology to allowing it to be used heavily. As we try to figure out how to teach this skill, we should be mindful of the fact that we were not very successful at teaching symbol manipulation before technology complicated the situation. It is tempting to gild the past, but weaknesses that we currently observe are probably not the result of technology. Our charge is to figure out how to fix them.

Besides acquiring skill in manipulating graphs, numerical data, and symbols, students need to be able to move easily between these representations. For example, given a straight line graph, students should be able to figure out its (approximate) equation. Given a table of data from an exponentially growing population, students should be able to figure out a formula for the function. Given a data set, students should be able to make a mental sketch of the data or match data with the correct sketch.

Mastery of the language of algebra requires a two-pronged approach:

1. What does it mean?
2. How do we do it?

Equal emphasis on both of these leads to students who know both what algebra means and how to use it correctly.

The future is likely to change the balance between these two because the skills required to do algebra are likely to change. However, we always will need to make sure students can use graphs, tables, symbols, and verbal descriptions fluently.

*Closing Remarks**

BEVERLY WILLIAMS: "Many firsts occurred over the last two days. I can relate what those firsts were for me personally. As a K to 12 math curriculum coordinator for Pulaski County School District in Arkansas, I experienced a first when I was asked to chair the task force that worked to put on this symposium, and I met Joan Ferrini-Mundy and other people at the Mathematical Sciences Education Board. And I had to learn about the National Research Council and the National Academy of Sciences. I think it strengthened me as a professional person to know what avenues of support are out there for me.

Another personal first—and hopefully it was for many of you, also—was the opportunity to interact with mathematicians these last two days. I interacted with many mathematics educators. I have grown as a person, and I thank you for the opportunity to be here and to share with mathematicians and with mathematics educators.

I hope many of you had firsts, that perhaps some of you who are collegiate people or from universities had an opportunity to sit beside an elementary-school teacher for the first time and talk about the role of algebra, for example. I remember yesterday morning when we were talking about curriculum, and someone said, 'We need to do this in the K to 6 grades.' And one of the K to 6 grade-level teachers said, 'I am that person. Tell me what you want me to do.' It was nice to have those interactions for the first time.

Another first happened this week that probably only a very few people are aware of. Actually, it happened yesterday: one of the participants told me last night that for the first time she felt the life of her child growing inside her. She said, 'I will never forget this symposium because the first kick I felt was sitting in this auditorium.' That may be a first for the Academy, also.

Martin van Reeuwijk and I were talking moments ago about the level of the people who are here and the continuum of knowledge present in this room. We talked about the hard questions, such as were our needs met, what were our goals, and did we have our own agenda for algebra. But I think the hardest part has yet to come. That will be when we go back to our states and regions and attempt to do a symposium and have a dialogue. We had a continuum with people here—possibly loaded at the front end with mathematics leaders—but when we get back to Arkansas, I think our continuum will be very long. We will have people who are on the leadership rolls of mathematics education and people we hope to bring into the continuum for the first time.

So, as you think about some of the dialogue that we have had the last two days, you should also think about the rich and different kinds of dialogue you are going to have back in your own states and how different the continuum will be.

*The following remarks by Beverly Williams were edited from verbatim transcriptions.

I heard a comment today—and I think it was very appropriate—about how we all need to be reminded as a community of mathematics educators that we cannot stand alone. It is imperative, for example, that we avoid the unintentional teacher bashing that we sometimes engage in with each other, where we start laying blame along the different levels, as in if only the universities did this or if the high schools did this or if the kindergartens... . We must remember our roles as leaders in the mathematics education communities and be really sensitive to what we say. The person who brought that up in one of the sessions today made a point on which we all can agree.

A few announcements. First, the Third International Mathematics and Science Study (TIMSS) resource kit will be mailed out to all the affiliated groups. If you don't get it in the next 12 months, call the NCTM headquarters office.

Also, the National Science Foundation is going to send all of the participants an updated listing of the National Center for Implementation of Standards-Based Mathematics Curricula, including contact information regarding NSF-funded curriculum projects. You should be receiving the proper information, with summaries and contact information.

In terms of regional efforts, I want to make sure that there is not a misconception about funding. The money for you to host a state or regional conference is whatever you can pull together. We at the National Council of Teachers of Mathematics will be glad to give you some hints of who to ask for funding. We know that many of you already have organized opportunities for dialogue at conferences and state meetings. Perhaps you can do a preconference on algebra to complement those efforts.

We greatly appreciate your being with us these last two days. Have a great summer and school year."

Appendices

Appendix A

The Nature and Role of Algebra in the K-14 Curriculum A National Symposium

AGENDA

Monday 26 May 1997
Capital Hilton

7-9 pm	Registration in the Lobby of the Hilton

Tuesday 27 May 1997
National Academy of Sciences (NAS)

8-8:15 am — Shuttle service departs from Hilton
8:15-9 am — Coffee and Registration continued at NAS

9-9:15 am — Welcome
Suzanne Woolsey
Chief Operating Officer, National Research Council

9:15-10:15 am — Opening Session
Hyman Bass and John Dossey

10:30 am-4:30 pm — Concurrent Sessions Auditorium, Lecture Room, AS 150

TOPIC #1
Algebra: A K-14 Curriculum Endeavor
Speakers: Andrew Gleason, James Kaput, Elizabeth Phillips
Facilitator: Sadie Bragg

TOPIC #2
Technology and Algebraic Reasoning
Speakers: Geoffrey Akst, Milton Norman, Rose Zbiek
Facilitator: Frank Demana

TOPIC #3
Algebra: What All Students Can Learn
Speakers: Jere Confrey, David Molina, Beatrice Moore-Harris
Facilitator: Irvin Vance

10:30-11:15 am	Concurrent Sessions, Rotation I
11:15 am-Noon	Questions and Response
Noon-1 pm	Lunch
1-1:45 pm	Concurrent Sessions, Rotation II
1:45-2:30 pm	Questions and Response Break
3-3:45 pm	Concurrent Sessions, Rotation III
3:45-4:30 pm	Questions and Response
4:30-5 pm	Exhibits, Great Hall
5-6pm	Reception, Great Hall

Shuttle Service to Hilton

Wednesday 28 May 1997
National Academy of Sciences

7-7:15 am	Shuttle service departs from Hilton
7:30-8 am	Coffee (exhibits continue throughout the day)
8-8:30 am	Synthesis of Day One, Auditorium Henry Pollak and Gail Burrill
8:45 am-12:15 pm	Algebra in the Curriculum
8:45-10:15 am	Participants Select One of the Following

Functions and Relations
Speakers: Philip Wagreich (elementary school),
Glenda Lappan (middle school), Christian Hirsch (high school)
Facilitator: Joan Ferrini-Mundy

Modeling
Speakers: Glenn Kleiman (middle school),
Landy Godbold (high school), Rosalie Dance (postsecondary)
Facilitator: Linda Rosen

10:45 am-12:15 pm Participants Select One of the Following

Structure
Speakers: Cornelia Tierney (elementary school),
Martin van Reeuwijk (middle school), Greg Foley (postsecondary)
Facilitator: Nancy Hawthorne

Language and Representation
Speakers: Rick Billstein (middle school), Diane Resek (high school), Deborah Hughes Hallet (postsecondary)
Facilitator: Skip Fennell

12:15-1:30 pm Lunch

1:30-2:30 pm Regional Planning Breakout Sessions

2:45-3:30 pm Next Steps
Closing Remarks

Shuttle Service to Hilton

Appendix B

*List of Participants**

Geoffrey Akst
Professor of Mathematics
Mathematics Department
Borough of Manhattan Community College/CUNY
199 Chambers Street
New York, NY 10007
212.346.8530 (W)
212.346.8550 (Fax)
akst@idt.net (E-mail)

Bruce Alberts
President
National Academy of Sciences
2101 Constitution Avenue, N.W.
Washington, DC 20418

Joel Albright
Mathematics Teacher
Douglas High School
Douglas School District
420 Patriot Drive
Box Elder, SD 57717
605.923.1464 (W)
202.358.3032 (Fax)
jalbrigh@hg.nasa.gov (E-mail)

John Andersen
Curriculum Director of Mathematics
Overland High School
12400 E. Jewell Avenue
Aurora, CO 80012
303.696.3749 (W)
303.696.3895 (Fax)
bfair@ooad.k12.us (E-mail)

Judy Anderson
Regional Program Director
California Mathematics Renaissance Program
2717 Lexington Place
Riverside, CA 92506
909.369.6434 (W)
judy_anderson@cams.edu (E-mail)

Angela Andrews
Scott School
1899 Carriage Hill Road
Lisle, IL 60532
630.420.6477 (W)
630.420-6471 (Fax)
angelaandrews@dcmrats.org (E-mail)

Charlie Aplin
Supervisor of Math and Science
Okaloosa County School Board
120 Lowery Place
Ft. Walton Beach, FL 32548
904.833.3170 (W)
904.833.3436 (Fax)
apline@mail.okaloosa.k12.fl.us (E-mail)

*The contact information provided here is current as of May 1997.

Joan Armistead
Senior Journal Editor
National Council of Teachers of Mathematics
1906 Association Drive
Reston, VA 20191-1593
703.620.9840 (W)
703.476.2970 (Fax)
jarmistead@nctm.org (E-mail)

Agnes Azzolino
Assistant Professor
Middlesex County College
Rt. 514 & Mill Road
Edison, NJ 08818-3050
908.739.3951 (W)
asquared@excaliber.com (E-mail)

Jane Barnard
Associate Professor of Mathematics
Armstrong Atlantic State University
University System of Georgia
11935 Abercorn Street
Savannah, GA 31419
912.921.2057 (W)
912.921.2083 (Fax)
jane_barnard@mailgate.armstrong.edu (E-mail)

Roy Barnes
Mathematics Supervisor
Arkansas Dept. of Education
Research and Design Team
#4 State Capitol Mall, Room 106A
Little Rock, AR 72201
501.682.5296 (W)
501.682.4886 (Fax)
barnesr@arkedu.K12.ar.us (E-mail)

Ann Bartosh
Mathematics Consultant
Kentucky Dept. of Education
500 Mero Street - 1836 CTP
Frankfort, KY 40601
502.564.2106 (W)
888.223.8387 (Fax)
abartosh@kde.state.ky.us (E-mail)

Hyman Bass
Professor of Mathematics
Columbia University
Mathematics Building, Room 509
116th Street at Broadway
New York, NY 10027
212.854.4114 (W)
212.854.8962 (Fax)
hb@math.columbia.edu (E-mail)

Jerry Becker
Director, National Council of
Teachers of Mathematics
Department of Curriculum and Instruction
Southern Illinois University
Carbondale, IL 62901-4610
618.453.4241 (W)
618.457.4244 (Fax)
jbecker@siu.edu (E-mail)

Nancy Berkas
Program Associate
Midwest Consortium for Mathematics
and Science Education
North Central Regional Educational Lab
1900 Spring Road, Suite 300
Oak Brook, IL 60521-1480
630.571.4700 (W)
630.571.4716 (Fax)

Rick Billstein
Professor of Mathematics
Mathematics Department
University of Montana
Missoula, MT 59812-1032
406.243.2603 (W)
406.243.2674 (Fax)
rickb@selway.umt.edu (E-mail)

Wesley Bird
Education Consultant
Missouri Dept. of Elementary and
Secondary Education
3301 S. Rangeline Road
Columbia, MO 65102
573.751.9069 (W)
573.526.7861 (Fax)

Rikki Blair
American Mathematical Association
of Two-Year Colleges
Lakeland Community College
7700 Clocktower Drive
Kirtland, OH 44094-5918
216.953.7341 (W)
rblair@lakeland.cc.oh.us (E-mail)

Art Block
Executive Group Director of Mathematics
Glencoe/McGraw-Hill
936 Eastwind Drive
Westerville, OH 43081
614.899.4457 (W)
614.899.4379 (Fax)

Debbie Blue
Associate Professor of Mathematics
Chair, Division of National Science and Mathematics
Oklahoma Baptist University
5 Chevy Chase
Shawnee, OK 74801
405.878.2029 (W)
405.878.2050 (Fax)
debbie_blue@mail.okbu.edu (E-mail)

L. Carey Bolster
Director, K-12 Math Projects
PBS MATHLINE
1320 Braddock Place
Alexandria, VA 22314
703.739.5402 (W)
703.739.7513 (Fax)
cbolster@pbs.org (E-mail)

Dee Bond
Mathematics Teacher
Overland High School
Cherry Creek School District
12400 East Jewell Avenue
Aurora, CO 80012
303.696.3700 (W)
303.696.3895 (Fax)
bondd@mscd.edu (E-mail)

Peg Bondorew
Mathematics Instructor and Director
CESAME
Northeastern University
716 Columbus Avenue, #378
Boston, MA 02120
617.373.8495 (W)
mbondorew@lynx.neu.edu (E-mail)

Alice Borja
Mathematics Specialist
Pacific Resources for Education and Learning
828 Fort Street Mall, Suite 500
Honolulu, HI 96913-4321
808.533.6000 (W)
808.533.7599 (Fax)
borja@prel.hawaii.edu (E-mail)

Laurie A. Boswell
Teacher, Profile Junior/Senior High School
Route 18
Bethlehem, NH 03574
603.823.7411 (W)
603.638.2802 (Fax)
lboswell@aol.com (E-mail)

Larry Bouldin
Professor of Mathematics and President
of the Tennessee Math Teachers Association
Roane State Community College
Patton Lane
Harriman, TN 37748
423.882.4553 (W)
423.354.3000 (Fax)
bouldin_cl@rscc.cc.tn.us (E-mail)

Spud Bradley
Program Officer
National Science Foundation
4201 Wilson Blvd.
Arlington, VA 22230
703.306.1628 (W)
703.306.0412 (Fax)
sbradley@nsf.gov (E-mail)

Sadie C. Bragg
Vice President of Academic Affairs
Borough of Manhattan Community College/CUNY
199 Chambers Street
New York, NY 10007
212.346.8820 (W)
212.346.8816 (Fax)
bmacdscb@cunyvm.cuny.edu (E-mail)

Daniel Brahier
Assistant Professor of Mathematics
Bowling Green State University
EDCI - 529 Education Building
Bowling Green, OH 43403
419.372.0339 (W)
419.372.7291 (Fax)
brahier@bgnet.bgsu.edu (E-mail)

Diane Briars
Director, National Council of
Teachers of Mathematics
Pittsburgh Public Schools
150 Maple Heights Road
Pittsburgh, PA 15232
412.338.8052 (W)

Nyla Bristow
Teacher, Centennial Elementary
Weld Co. School District 6
1400 37th Street
Greeley, CO
970.339.5320 (W)
970.330.7193 (Fax)
nylab@psd.k12.co.us (E-mail)

Cathy Brown
Mathematics Assessment Specialist
Oregon Department of Education
Office of Assessment and Evaluation
255 Capitol Street, N.E.
Salem, OR 97310-0203
503.378.5585 (W)
503.373.7968 (Fax)
cathy.brown@state.or.us (E-mail)

Blanche Brownley
Team Leader
MSTI Senior Learning Years
DC Public Schools
Rabaut Conference & Training Center
100 Peabody St., N.W., Room 310
Washington, DC 20011
202.544.6020 (W)
202.544.6017 (Fax)
brownley_b@mercury.k12.dc.us (E-mail)

Douglas Brumbaugh
Professor, College of Education
University of Central Florida
Orlando, FL 32816-1250
407.823.2045 (W)
407.823.5964 (Fax)

Maurice Burke
Co-Director
Systemic Initiative for Montana
Mathematics and Science
Department of Mathematical Sciences
Montana State University
Bozeman, MT 59717-0240
406.994.5344 (W)
burke@mathfs.math.montana.edu (E-mail)

Gail F. Burrill
President, National Council of
Teachers of Mathematics
Center for Research in Mathematics and Science
12155 West Luther Avenue
Hales Corners, WI 53130
414.425.7049 (W)
414.425.6442 (Fax)
gburrill@macc.wisc.edu (E-mail)

Larry Campbell
Associate Professor of Mathematics
Southwest Missouri State University
901 South National
Springfield, MO 65804
417.836.5942 (W)
417.836.5610 (Fax)
lnc999f@wpgate.smsu.edu (E-mail)

James Cangelosi
Mathematics Professor
Mathematics and Statistics Department
Utah State University
Logan, UT 84322-3900
801.797.1415 (W)
801.797.1822 (Fax)
jcang@math.usu.edu (E-mail)

Ann Carlyle
Director, National Council of
Teachers of Mathematics
Ellwood Elementary
P.O. Box 1313
Goleta, CA 93116
805.964.1616 (W)
805.964.1072 (Fax)
acarly@ellwood.sbceo.k12.ca.us (E-mail)

Betty Causey-Lee
Supervisor, Detroit Systemic Initiative
Schools Center Building
5057 Woodward Avenue, Room 932
Detroit, MI 48202
313.494.1616 (W)

Kathleen Chadwick
Middle School Teacher
Exeter-West Greenwich Jr. High
930 Nooseneck Hill Road
West Greenwich, RI 02879
401.397.6898 (W)
Chadwickk@aol.com (E-mail)

Alverna Champion
Program Officer
National Science Foundation
4201 Wilson Blvd.
Arlington, VA 22230
703.306.1620 (W)
703.306.0412 (Fax)
achampio@nsf.gov (E-mail)

Pamela Chandler
Mathematics Coordinator
Fort Bend School District
16431 Lexington Blvd.
Sugar Land, TX 77479
281.634.1253 (W)
281.634.1713 (Fax)
pchan@tenet.edu (E-mail)

Kathleen Chapman
Journal Editor
National Council of Teachers of Mathematics
1906 Association Drive
Reston, VA 20191-1593
703.620.9840 (W)
703.476.2970 (Fax)
kchapman@nctm.org (E-mail)

Bruce Chester
6408 Westward Drive
Loveland, OH 45140
513.729.0698 (W)

Leigh Childs
Mathematics Consultant
San Diego County Office of Education
2615 Loring Street
San Diego, CA 92109
619.273.1421 (W)
619.273.6541 (Fax)
leigh@sdcpe.k12.ca.us (E-mail)
leigh_childs@cams.edu (E-mail)

Douglas Clements
Professor, Department of Learning
and Instruction
State University of New York at Buffalo
593 Baldy Hall
Buffalo, NY 14260
716.645.3158 (W)
716.645.3161 (Fax)
clements@acsu.buffalo.edu (E-mail)

Tom Clonts
7132 North 12th Place
Phoenix, AZ 85020
602.271.2400 (W)

Karen Cockburn
Mathematics Department Chair
Joel E. Ferris High School
Spokane School District 81
3020 E. 37th
Spokane, WA 99223
509.353.4400 (W)
509.358.4669 (Fax)
kcockburn@aol.com (E-mail)

Terry Coes
Director, National Council of
Teachers of Mathematics
Rocky Hill School
Wakefield, RI 02879
401.884.9070 (W)
401.885.4985 (Fax)
lcoes@aol.com (E-mail)

Jere Confrey
Professor of Mathematics Education
University of Texas at Austin
Austin, TX 78712-1294
512.471.3747 (W)
512.471.8857 (Fax)
jere@mail.utexas.edu (E-mail)

Ruth Cossey
Mills College
5000 McArthur Boulevard
Oakland, CA 94613
510.430.3159 (W)
510.430.3379 (Fax)
rcossey@mills.edu (E-mail)

Shari Wilson Coston
Executive Director
Arkansas Educational Renewal Consortium
Henderson State University
P.O. Box 7894 HSU
Arkadelphia, AR 71998-0001
501.230.5353 (W)
501.230.5459 (Fax)
mcoston@direclynx.net (E-mail)

Joan Cox
Mathematics Facilitator
Memphis Systemic Initiative
2385 Central
Memphis, TN 38104
901.722.4552 (W)
901.722.4557 (Fax)

Elaine Crawford
Mathematics Facilitator
for High-School Assessment
Maryland State Department of Education
200 W. Baltimore Street
Baltimore, MD 21201
410.767.0328 (W)
410.333.2379 (Fax)
elainecraw@aol.com (E-mail)

Rosalie Dance
7406 Jackson Avenue
Takoma Park, MD 20912
301.270.0086 (W)
202.687.6067 (Fax)
dancer@gusun.acc.georgetown.edu (E-mail)

Thomas DeFranco
Associate Professor
University of Connecticut
U-33; 249 Glenbrook Road
Storrs, CT 06269
860.486.0288 (W)
860.486.0210 (Fax)
defrano@uconnvm.uconn.edu (E-mail)

Franklin Demana
Professor Emeritus
Department of Mathematics
Ohio State University
231 W. 18th Avenue
Columbus, OH 43210
614.292.1934 (W)
demanaf@math.ohio-state.edu (E-mail)

Robert L. Devaney
Professor of Mathematics
Department of Mathematics
Boston University
111 Cummington Street
Boston, MA 02215
617.353.4560 (W)
617.353.8100 (Fax)
bob@math.bu.edu (E-mail)

Wendy Diskin
Sixth-grade Teacher
Mayer Elementary School
Mayer Unified School District
17300 Mule Deer Drive
Mayer, AZ 86333
520.632.9679 (W)
520.632.9610 (Fax)
wendy@northlink.com (E-mail)

Juli Dixon
Assistant Professor
University of Nevada
Box 453005
Las Vegas, NV
702.895.1448 (W)
702.895.4878 (Fax)
dixon@nevada.edu (E-mail)

Lori Donoho
Program Specialist
Eisenhower Consortium for Mathematics
and Science Education at SERVE
41 Marietta Street, N.W., Suite 1000
Atlanta, GA 30312
404.893.0114 (W)

John A. Dossey
Department of Mathematics
Illinois State University
4520 Mathematics
Normal, IL 61790-4520
309.438.7707 (W)
309.438.5866 (Fax)
jdossey@math.ilstu.edu (E-mail)

Linda Dritsas
Mathematics Coordinator
Fresno Systemic Initiative
IMC, Bldg. 3
3132 E. Fairmont Avenue
Fresno, CA 93726
209.441.3642 (W)
209.265.2749 (Fax)

Jim Dunlap
1900 Calafia Street
Glendale, CA 91208
213.625.4209 (W)

Joyce Eaton
Teacher, St. Louis Public Schools
450 Des Peres
St. Louis, MO 63110
314.863.7266 (W)
314.863.4638 (Fax)

Sue Eddins
Illinois Mathematics and Science Academy
1500 West Sullivan Road
Aurora, IL 60506-1000
630.907.5966 (W)
630.907.5976 (Fax)
eddins@imsa.edu (E-mail)

Marjorie Enneking
Professor of Mathematics
Department of Mathematics
Portland State University
Portland, OR 97207-0751
503.725.3643 (W)
503.725.3661 (Fax)
marj@mth.pdx.edu (E-mail)

Edward Esty
Staff Consultant
Commission on the Future of the Standards
4608 Drummond Avenue
Chevy Chase, MD 20815-5433
301.656.7274 (W)
eesty@aol.com (E-mail)

Jerry Evans
Mathematics Specialist
Utah State Office of Education
250 East 500 South
Salt Lake City, UT 84111
801.538.7794 (W)
801.838.7769 (Fax)
jerry.evans@usoe.kiz.ut.us (E-mail)

Joyce Evans
National Science Foundation
4201 Wilson Blvd.
Arlington, VA 22230
703.306.1628 (W)
703.306.0412 (Fax)
jevans@nsf.gov (E-mail)

Richard Evans
Professor, Department of Mathematics
Hyde Hall
Plymouth State College
Plymouth, NH 03264
603.535.2487 (W)
603.535.2943 (Fax)
evans@oz.pltymouth.edu (E-mail)

Skip Fennell
Division of Elementary, Secondary,
and Informal Education
National Science Foundation
4201 Wilson Boulevard
Room 885.33
Arlington, VA 22230
703.306.6835 (W)
703.306.0412 (Fax)
ffennell@nsf.gov (E-mail)

Shelley Ferguson
California Mathematics Project Coordinator
California Mathematics Project
6475 Alvarado Road, Suite 206
San Diego, CA 92120-5006
619.594.6416 (W)
619.594.1581 (Fax)
sfergu@sunstroke.sdsu.edu (E-mail)

Henry Ferrer
District Director
Miami Systemic Initiative
1500 Biscayne Blvd., Suite 327
Miami, FL 33132
305.995.2341 (W)
305.995.1916 (Fax)
hferrer@sbab.dade.k12.fl.us (E-mail)

Joan Ferrini-Mundy
Mathematical Sciences Education Board
National Research Council
2101 Constitution Avenue, N.W.
Washington, DC 20418
202.334.1472 (W)
202.334.1453 (Fax)
jferrini@nas.edu (E-mail)

Mary E. Ferris
Affiliated Group Services Staff
National Council of Teachers of Mathematics
1906 Association Drive
Reston, VA 20191-1593
703.620.9840 (W)
703.476.2970 (Fax)
mferris@nctm.org (E-mail)

Beverly Ferrucci
Chair, Mathematics Department
Keene State College
229 Main Street
Keene, NH 03435-2001
603.358.2506 (W)
603.358.2897 (Fax)
bferrucc@keene.edu (E-mail)

Carol Findell
Professor and Program Coordinator
Mathematics Education
School of Education
Boston University
605 Commonwealth Avenue
Boston, MA 02215
617.353.4226 (W)
617.353.2909 (Fax)

Bradford Findell
Mathematical Sciences Education Board
National Research Council
2101 Constitution Avenue, N.W.
Washington, DC 20418
202.334.1472 (W)
202.334.1453 (Fax)
bfindell@nas.edu (E-mail)

Alfinio Flores
Associate Professor
Arizona State University
Division of Curriculum and Instruction
Tempe, AZ 85287-0911
602.965.3101 (W)
602.965.6604 (Fax)
alfinio@asu.edu (E-mail)

Gregory D. Foley
Division of Mathematical and Information Science
Sam Houston State University
Box 2206
Huntsville, TX 77341-2206
409.294.3708 (W)
409.294.1882 (Fax)
mth_gdf@shsu.edu (E-mail)

Catherine Franklin
Mathematical Supervisor and
Program Coordinator
Muscogee County Schools
Columbus State University
Room 102 Stanley Hall; CRMC
4225 University Avenue
Columbus, GA 31907
706.569.3146 (W)
706.562.1477 (Fax)
franklin_catherine@colstate.edu (E-mail)

Jennifer Funderburk
Mathematics Teacher
Colquitt County High School
Colquitt County Schools
1800 Park Avenue
Moultrie, GA 31768
912.890.6181 (W)
912.890.6166 (Fax)

Dianne Gann
Sixth-grade Teacher
Root Elementary School
Fayetteville School District
1529 Mission Boulevard
Fayetteville, AR 72701
501.444.3075 (W)
dgann@nwsc.k12.ar.us (E-mail)

Robert Garvey
Mathematics Chairman
Independent Louisville
Collegiate School
2427 Glenmary Avenue
Louisville, KY 40204
502.451.5330 (W)
502.451.5591 (Fax)
bgarvey@aol.com (E-mail)

James Gates
Mathematical Sciences Education Board
National Research Council
2101 Constitution Avenue, N.W.
Washington, DC 20418
202.334.1472 (W)
202.334.1453 (Fax)
jgates@nas.edu (E-mail)

Ladnor Geissinger
Mathematics Chair
Institute for Academic Technology
University of North Carolina
2525 Meridian Parkway, Suite 400
Durham, NC 27713
919.405.1925 (W)
919.560.5047 (Fax)
ladnor.let@mhs.unc.edu (E-mail)

Brenda Gentry-Norton
Mathematics Consultant
WestEd Eisenhower Regional Consortium
for Science and Mathematics Education
c/o Ann Muench
12345 El Monte Drive
Los Altos Hills, CA 94022
415.241.2730 (W)
415.241.2746 (Fax)

William Geppert
Mathematics Staff Associate
Mid-Atlantic Eisenhower Regional Consortium
for Mathematics and Science Education
Research for Better Schools
444 North Third Street
Philadelphia, PA 19123-4107
215.574.9300 (W)
215.574.0133 (Fax)

Andy Gleason
110 Larchwood Drive
Cambridge, MA 02138
617.864.5095 (W)
gleason@math.harvard.edu (E-mail)

Landy Godbold
Math Teacher
The Westminster Schools
1424 W. Paces Ferry Road NW
Atlanta, GA 30327
404.355.8673 (W)
404.355.6606 (Fax)
L.godbold@mail.comap.com (E-mail)

Jean Gomer
Mathematics Teacher
Department Head
Deubrook Area School 5-6
Deubrook Area High School
Box 346
White, SD 57276
605.629.3201 (W)
605.629.3701 (Fax)
trewatha@itctel.com (E-mail)

Danny Goroff
Office of Science and Technology Policy
Executive Office of the President
OEOB #436
Washington, DC 20502
202.456.6003 (W)
dgoroff@ostp.eop.gov (E-mail)

Sandra Gothard
Teacher
Carey Junior High
Laramie County School District #1
1780 E. Pershing Blvd.
Cheyenne, WY 82001
307.771.2580 (W)
307.771.2578 (Fax)
goth0100@spacelink.msfc.nasa.gov (E-mail)

Joan Grampp
Mathematics/Science Resource Teacher
Jackie Robinson Middle School
and South Division High School
Milwaukee Public Schools
3245 N. 37th
Milwaukee, WI 53216
414.438.3694 (W)
414.4387.3470 (Fax)

Linda Griffin
Associate Professor of Mathematics Education
University of Central AR- Conway
201 Donaghey Avenue - UCA Box 4912
Conway, AR 72035-0001
501.450.5663 (W)
501.450.5009 (Fax)
lindag@ccl.vca.edu (E-mail)

Ena Gross
Associate Professor
Division of Teacher Education
Virginia Commonwealth University
1015 West Main Street
P.O. Box 842020
Richmond, VA 23284-2020
804.828.1305 (W)
804.828.1323 (Fax)
egross@atlas.vcu.edu (E-mail)

Barbara Guiduli
Mathematics Teacher
Burlington High School
52 Institute Road
Burlington, VT 05401
802.864.8410 (W)
rguidli@zoo.ovm.edu (E-mail)

Bill Hadley
1308 Milton Street
Pittsburgh, PA 15218
412.260.2420 (W)
412.268.1266 (Fax)
bill_hadley@cmu.edu (E-mail)

Marilyn L. Hala
Director, Professional Programs
National Council of Teachers
of Mathematics
1906 Association Drive
Reston, VA 20191
703.620.9840 (W)
703.476.2970 (Fax)
mhala@nctm.org (E-mail)

Basia Hall
Teacher, Alief Hastings High School (North)
Alief ISD
P.O. Box 68
Alief, TX 77411
281.498.8110 (W)
basia@tenet.edu (E-mail)

Lori Hamada
Mathematics Teacher
Kingsburg High School
Kingsburg School District
1900 18th Avenue
Kingsburg, CA 93631
209.497.3903 (W)
209.225.7801 (Fax)
hamada@cpm.org (E-mail)

Randy Harter
Mathematics Specialist
Buncombe County Schools
175 Bingham Road
Asheville, NC 28806
704.255.5956 (W)
704.255.5944 (Fax)

Charlotte Hartman
Administrator for Curricular Development
Vancouver School District No 37
P.O. Box 8937
605 North Devine Rd.
Vancouver, WA 98668-8937
360.696.7011 (W)
360.696.7027 (Fax)
chartman@vannet.k12.wa.us (E-mail)

Jim Hassel
Mathematics Teacher
Berea High School
165 E. Bagley Road
Berea, OH 44017
216.234.5419 (W)
216.826.6973 (Fax)
jhassel@baldwinw.edu (E-mail)

Nancy Hawthorne
Manager, Affiliated Group Services
National Council of Teachers
of Mathematics
1906 Association Drive
Reston, VA 20191
703.620.9840 (W)
703.476.2970 (Fax)
nhawthorne@nctm.org (E-mail)

Claire Heidema
Staff Associate
Eisenhower High Plains Consortium
for Mathematics and Science at McREL
2550 South Parker Road, Suite 500
Aurora, CO 80014
800.949.6387 (W)
303.337.3005 (Fax)

Terese Herrera
Mathematics Resource Specialist
Eisenhower National Clearinghouse
for Mathematics and Science Education
The Ohio State University
1929 Kenny Road
Columbus, OH 43210-1079

Jim Hersberger
Professor of Mathematics
Indiana University-Purdue University
Fort Wayne (IPFW)
2101 Coliseum Blvd.
Fort Wayne, IN 46805
219.432.7077 (W)
hersberg@smtplink.ipfw.indiana.edu. (E-mail)

Margie Hill
Mathematics Supervisor
Blue Valley School District Office
15020 Metcalf
Overland Park, KS 66223
913.681.4059 (W)
913.681.4075 (Fax)
marghill@unicom.net (E-mail)

Christian Hirsch
Professor of Mathematics
and Mathematics Education
Department of Mathematics and Statistics
Western Michigan University
Kalamazoo, MI 49009
616.387.4526 (W)
616.387.4530 (Fax)
christian.hirsch@wmich.edu (E-mail)

Kathleen Hollowell
Director, Mathematics and Science
University of Delaware
105 Pearson Hall
Newark, DE 19716
302.831.1022 (W)
302.831.3137 (Fax)
kathyh@udel.edu (E-mail)

Ellen Hook
Mathematics Teacher
Granby High School
Norfolk Public Schools
7101 Granby Street
Norfolk, VA 23505
757.441.1265 (W)
757.518.0202 (Fax)
ehook@pen.k12.va.us (E-mail)

Bill Hopkins
Director of Mathematics
Texas Education Agency
1701 North Congress Avenue
Austin, TX 78701
512.463.9585 (W)
512.463.8057 (Fax)
bhopkin@tenet.edu (E-mail)

Susanne Hossler
Mathematics Teacher
Hyde Park Academy of Science
and Mathematics
Clark County School District
900 Hinson Street
Las Vegas, NV 89107
702.799.4260 (W)
suzagirls@aol.com (E-mail)

Peggy House
Director, National Council
of Teachers of Mathematics
Northern Michigan University
Seaborg Center
1401 Presque Isle Avenue
Marquette, MI 49855
906.227.2002 (W)
906.227.2013 (Fax)
phouse@nmu.edu (E-mail)

Jean Howard
C.R. Anderson Middle School
985 Terrence Road
Helena, MT 59602-9359
406.447.8910 (W)
jkhoward@metnet.mt.gov (E-mail)

Roger Howe
Professor of Mathematics
Department of Mathematics
Yale University
P.O. Box 208283
New Haven, CT 06520-8283
203.432.4686 (W)
203.432.7316 (Fax)
howe@math.yale.edu (E-mail)

Deborah Hughes Hallet
Professor of Mathematics
Mathematics Department
Harvard University
Cambridge, MA 02138
617.495.5358 (W)
617.576.6967 (Fax)
dhh@math.harvard.edu (E-mail)

Gerunda Hughes
Assistant Professor of Mathematics Education
Howard University
2441 4th St., N.W.
Washington, DC 20059
202.806.7343 (W)
202.806.5297 (Fax)
ghughes@fac.howard.edu (E-mail)

William Hunt
Systems Coordinator
Department of Education and Allied Studies
John Carroll University
University Heights, OH 44118
216.397.1714 (W)
216.238.3147 (Fax)
bhunt@mcs.kent.edu (E-mail)

Linda Hutchison
Assistant Professor of Mathematics Education
University of Wyoming
P.O. Box 3992
Laramie, WY 82071-3992
307.766.6852 (W)
307.766.3792 (Fax)
1hutch@uwyo.edu (E-mail)

Tom Hymers
K-12 Instructional Coordinator, Mathematics
Naperville North High School
Naperville School District #203
899 N. Mill Street
Naperville, IL 60563-8998
630.420.6513 (W)
630.420.2554 (Fax)

Ramona Irvin
Mathematical Sciences
Education Board Staff
National Research Council
2101 Constitution Avenue, N.W.
Washington, DC 20418
202.334.1472 (W)
202.334.1453 (Fax)
rirvin@nas.edu (E-mail)

Beryl Jackson
Director, Middle School Math Project
PBS MATHLINE
1320 Braddock Place
Alexandria, VA 22314
703.739.5124 (W)
703.739.7513 (Fax)
bjackson@pbs.org (E-mail)

Judith Jacobs
Immediate Past President
California State Polytechnic University
3801 West Temple Avenue
Pomona, CA 91768
909.869.3973 (W)
909.869.4616 (Fax)
jejacobs@csupomona.edu (E-mail)

Rick Jennings
Teacher, Yakima School District
702 South 40th Avenue
Yakima, WA 98908
509.965.1649 (W)
509.966.8933 (Fax)
rickjenn@msn.com (E-mail)

Carolyn Johnson
Project SUCCESS
2401 Hunting Horn Way
Virginia Beach, VA 23456

Darlene Johnson
Mathematics Teacher
and Department Chair
Altus High School
Altus Public Schools
400 N. Park Avenue
Altus, OK 73521
405.481.2167 (W)
405.481.2545 (Fax)
park@intellisys.net (E-mail)

W. Tad Johnston
Mathematics Facilitator
Maine Mathematics and Science Alliance
P.O. Box 5359
Augusta, ME 04332
207.288.2802 (W)
207.288.5071 (Fax)
tjohnsto@mdibs.u98.k12.me.us (E-mail)

Chancey Jones
Principal Measurement Specialist
Educational Testing Service
Rosedale Road
Princeton, NJ 08541
609.683.2640 (W)
609.734.1909 (Fax)

Patricia Jones
Assistant Professor, Mathematics
University of Southwestern Louisiana
P.O. Box 41010
Lafayette, LA 70504
318.482.5293 (W)
318.482.5346 (Fax)
plj0786@usl.edu (E-mail)

Jeane Joyner
Chair, North Carolina Department
of Public Instruction
3021 Eton Road
Raleigh, NC 27608
919.715.1864 (W)
919.715.1841 (Fax)
jjoyner@dpistate.nc.us (E-mail)

James J. Kaput
Department of Mathematics
University of Massachusetts at Dartmouth
North Dartmouth, MA 02747
508.999.8321 (W)
508.999.8901 or 508.910-6917 (Fax)
Jkaput@umassd.edu (E-mail)

Hank Kepner
Program Officer
National Science Foundation
4201 Wilson Blvd.
Arlington, VA 22230
414.229.4844 (W)
414.229.5571 (Fax)
kepner@csd.uwm.edu (E-mail)

Rob Kimball
Head, Math and Physics Department
Wake Community College
P.O. Box 6
Knightsdale, NC 27545
919.662.3602 (W)

Jean Klanica
Director, IMP-RMR
Eaglecrest High School
Cherry Creek School District
1500 S. Picadilly Street
Aurora, CO 80015
303.751.0895 (W)
303.338.4301 (Fax)
jklanica@csn.net (E-mail)

Glenn Kleiman
Vice President and Senior Scientist
Education Development Center, Inc.
55 Chapel Street
Newton, MA 02158
617.969.7100 (W)
GlennK@edc.org (E-mail)

Mary Ellen Komorowski
Professor of Mathematics
and Marshall County School Board Member
Mathematics Department
Bethany College
107 Richardson Hall
Bethany, WV 26032
304.829.7742 (W)
m.komorowski@mail.bethanywv.edu (E-mail)

Roberta Koss
Redwood High School
71 Twelveoak Hill Drive
San Rafael, CA 94903
415.945.3600 (W)

Martha Kuntz
Teacher and VP of Middle School SCCTM
Alice Birney Middle School
7750 Pinehurst Street
Charleston, SC 29420
803.764.2212 (W)

Carole Lacampagne
U.S. Department of Education
400 Maryland Avenue, SW
Washington, DC 20202
202.708.5366 (W)

Larry Lamb
Mathematics Coordinator
West Virginia Department of Education
1900 Kanawha Blvd. E., Bldg. 6, Rm. 330
Charleston, WV 25315
304.558.7805 (W)
304.558.0459 (Fax)
llamb@access.k12.wv.us (E-mail)

Diana Lambdin
Assistant Professor
School of Education
Indiana University at Bloomington
201 North Rose Street
Bloomington, IN 47405
812.856.8149 (W)
812.856.8440 (Fax)
lambdin@indiana.edu (E-mail)

Ron Lancaster
St. Mildred's Lightbourn School
35 Haddon Avenue North
Hamilton, Ontario Canada l8S 4A2
905.845.5830 (W)
905.845.4799 (Fax)
ron2718@netaccess.on.ca (E-mail)

Glenda Lappan
Professor of Mathematics
Michigan State University
101 Will House
East Lansing, MI 48824
517.353.4657 (W)
517.432.2872 (Fax)
glappan@math.msu.edu (E-mail)

Christine Larson
Assistant Professor
South Dakota State University
1423 8th Street, S.
Brookings, SD 57006
605.688.6219 (W)
605.688.5880 (Fax)
ma11@sdsumus.sdstate.edu (E-mail)

Sylvia Lazarnick
Mathematics Teacher
Bellows Free Academy
71 South Main Street
St. Albans, VT 05478
802.527.7576 (W)

Ellen Lee
Mathematics Department Chairman
St. Joseph's Academy
Diocese of Baton Rouge
3015 Broussard St.
Baton Rouge, LA 70808
504.383.7207 (W)
504.344.5714 (Fax)

Steve Leinwand
Director, National Council of
Teachers of Mathematics
142 Tavern Circle
Middletown, CT 06457
860.566.2645 (W)
860.566.5623 (Fax)
stevemath@aol.com (E-mail)

Larry Leutzinger
Professor, Mathematics Department
Wright Hall
University of Northern Iowa
Cedar Falls, IA 50614
319.273.6958 (W)
319.273.2546 (Fax)
leutzing@uni.edu (E-mail)

Jim Lightbourne
Program Officer
National Science Foundation
4201 Wilson Blvd., Room 835
Arlington, VA 22230
703.306.1665 (W)
703.306.0445 (Fax)
jhlightb@nsf.gov (E-mail)

Ken Lloyd
President, PCTM
Mathematics Teacher
West Scranton High School
1201 Luzerne Street
Scranton, PA 18504
717.348.3647 (W)
717.348.3594 (Fax)
knwld@aol.com (E-mail)

Fred Longhart
Mathematics Teacher
Flathead High School
School District #5 Kalispell
644 4th Avenue, W.
Kalispell, MT 59901
406.756.5099 (W)
406.756.4595 (Fax)
longhart@digisys.net (E-mail)

Johnny Lott
Director, National Council
of Teachers of Mathematics
Department of Math Sciences
University of Montana
Missoula, MT 59812
406.243.2696 (W)
406.243.2674 (Fax)
ma_jw@selway.umt.edu (E-mail)

Andre Lubecke
Associate Professor
Lander University
108 Loblolly Circle
Greenwood, SC 29649-1138
864.388.8212 (W)
864.388.8126 (Fax)
alubercke@lander.edu (E-mail)

Renee Lundy
Math Manager
Teachers Academy
of Mathematics and Science
3424 S. State Street
Chicago, IL 60616
312.949.2550 (W)
312.808.9257 (Fax)
rlundy@tams.iit.edu (E-mail)

Carol Malloy
School of Education
University of North Carolina at Chapel Hill
1125 Anderson Street
Durham, NC 27705
919.962.6607 (W)
919.962.1533 (Fax)
cmalloy@email.unc.edu (E-mail)

Alfred Manaster
Department of Mathematics
University of California, San Diego
9500 Gilman Drive
La Jolla, CA 92093-0112
619.534.2644 (W)
619.534.5273 (Fax)
amanaster@ucsd.edu (E-mail)

Melissa Manzano
Hubbard Heights Elementary School
7484 Tallow Wind, Apt. H
Ft. Worth, TX 76133
817.922.6580 (W)
817.922.6986 (Fax)
melmanzano@aol.com (E-mail)

John Martin
Chairman and Professor
Department of Mathematics and Computer Science
Shippensburg University
Shippensburg, PA 17257
717.532.1781 (W)
jfmart@ark.ship.edu (E-mail)

Rafael Martinez
Division Director, ESIE
National Science Foundation
4201 Wilson Blvd.
Arlington, VA 22230
703.306.1665 (W)
703.306.0445 (Fax)

Jim Matthews
Professor of Mathematics
and Computer Science
Mathematics Department
Siena College
Loudenville, NY 12211
518.783.2466 (W)
518.783.2986 (Fax)
matthews@siena.edu (E-mail)

Pamela Matthews
Instructor, Mathematics and Statistics Department
American University
4400 Massachusetts Avenue, N.W.
Washington, DC 20016
202.885.3143 (W)
202.885.3155 (Fax)
pmatthe@american.edu (E-mail)

Linda McCarthy
Teacher, Ichabod Crane High School
Route 9
Valatie, NY 12184
518.758.7577 (W)

Gail McClure
Teacher, DuPont Junior High
Kanawha County Schools
201 West Main Street
Belle, WV 25015
304.348.1977 (W)
gmcclure@access.k12.wv.us (E-mail)

Jean McGehee
Mathematics Professor
Department of Mathematics
Northern Arizona University
P.O. Box 5717
Flagstaff, AZ 86011
520.523.6862 (W)
520.523.5847 (Fax)
jean.mcgehee@nav.edu (E-mail)

Gloria McRae
Mathematics Instructor
East Central Community College
P.O. Box 129
Decatur, MS 39327
601.635.2111 (W)
601.267.9665 (Fax)
mcraegs@netdoor.com (E-mail)

Efrain Melendez
Resource Teacher
Los Angeles Systemic Initiative
450 N. Grand Avenue, Room A819
Los Angeles, CA 90012
213.625.6421 (W)
213.626.7785 (Fax)

Carol Midgett
Southport Elementary
1002 Bussells Turn
Southport, NC 28461
910.457.6036 (W)
910.457.6042 (Fax)

Clarence Miller
National Science Foundation
3512 North Calvert Street
Baltimore, MD 21218
202.208.3896 (W)

Susan Miller
Mathematics Chairman
Carver High School of Engineering and Science
Philadelphia Systemic Initiative
17th and Norris Streets
Philadelphia, PA 19121
215.684.5079 (W)
215.684.5151 (Fax)

Mary Mitchell
Supervisor of Mathematics K-12
Trenton Public Schools
Central Services Building
108 North Clinton Avenue
Trenton, NJ 08609
609.278.3015 (W)
609.278.3001 (Fax)

David D. Molina
Charles A. Dana Center
University of Texas at Austin
ECN 3.200 2901 North IH 35
Austin, TX 78722
512.471.6190 (W)
dmolina@mail.utexas.edu (E-mail)

Beatrice Moore-Harris
4710 Sabrina
Houston, TX 77066
713.748.6033 (W)
713.746.5283 (Fax)

Kathy Mowers
Professor, Owensboro Community College
4800 New Hartford Road
Owensboro, KY 42303
502.686.4564 (W)
502.686.4550 (Fax)
kamowe0@pop.uky.edu (E-mail)

John Moyer
Professor of Mathematics
Department of Mathematics,
Statistics, and Computer Science
Marquette University
P.O. Box 1881
Milwaukee, WI 53201-1881
414.288.5299 (W)
414.288.5472 (Fax)
johnm@mscs.mu.edu (E-mail)

Tom Muchlinski
Mathematics Teacher
Watzata
305 Vicksburg Lane
Plymouth, MN 55447
612.476.3035 (W)
612.476.3289 (Fax)
tmuchlin@informns.k12mn.us (E-mail)

Mari Muri
Mathematics Consultant
Connecticut Department of Education
165 Capitol Avenue
Hartford, CT 06106
860.566.4588 (W)
860.566.5623 (Fax)
mmuri@wesleyan.edu (E-mail)

Throys Nachlinger
2800 West Semmes
Osceola, AR 72370
870.563.2192 (W)
870.563.6732 (Fax)

Susan Nancarrow
Teacher, West Seaford Elementary
511 Sussex Avenue
Seaford, DE 19973
302.629.9352 (W)
nancar@udel.edu (E-mail)

Gary Nanke
Mathematics Consultant
AEA7 & Pricelob School
3712 Cedar Heights Drive
Cedar Falls, IA 50613
319.273.8215 (W)
319.273.8219 (Fax)
gnanka@aea7.k12.ia.us (E-mail)

Jan Niemira
Teacher, Mattawan Regional High School
Mattawan Aberdeen Regional School District
Atlantic Avenue
Mattawan, NJ 07747
908.290.2841 (W)
jniemira@exit109.com (E-mail)

Irene Norde
Supervisor, Mathematics
Detroit Systemic Initiative
18945 Parkside
Detroit, MI 48221
313.494.1612 (W)
313.494.7864 (Fax)
inorde@sun.science.wayne.edu (E-mail)

Milton O. Norman
Teacher, 9-12
Norfolk Public Schools
7101 Granby St.
Norfolk, VA 23505
757.441.1265 (W)
757.486.8496 (Fax)
mnorman@pen.k12.va.us (E-mail)

Theresa Frazier Norris
Cincinnati PS/Rm. 114
P.O. Box 5381
Cincinnati, OH 45201-5381

Denise Obert
High-School Mathematics Educator
Skowhegan Area High School
Maine School Administrative District No 54
Skowhegan, ME
207.474.5511 (W)
207.474.0992 (Fax)

Melfried Olson
Professor of Mathematics
Western Illinois University
1 University Circle
Macomb, IL 61455
309.298.2275 (W)
309.298.2585 (Fax)
olsonm@ccmail.wiu.bgu.edu (E-mail)

Barbara Pence
AMTE Representative
San Jose State University
10639 Ridgeview Way
San Jose, CA 95127
408.924.5142 (W)
408.924.5080 (Fax)
pence@sjsumes.sjsu.edu (E-mail)

Betty Phillips
Senior Academic Specialist
Department of Mathematics
Michigan State University
East Lansing, MI 48824
517.353.3835 (W)
517.432.2872 (Fax)
ephillips@math.msu.edu (E-mail)

Debbie Phillips
Einstein Fellow
U.S. Department of Education
130 E. Glebe Road
Alexandria, VA 22305
202.260.2654 (W)
202.205.0302 (Fax)
debbie_phillips@ed.gov (E-mail)

Anthony Piccolino
Professor, Montclair State University
Valley Road
Upper Montclair, NJ 07043
201.655.7247 (W)
201.655.7686 (Fax)
piccolino@alpha.montclair.edu (E-mail)

Gary Plummer
Math Resources Teacher
Philadelphia Systemic Initiative
7950 Temple Road
Philadelphia, PA 19150
215.276.5252 (W)
215.276.5849 (Fax)
gplummer@mail.idt.net (E-mail)

Henry Pollak
40 Edgewood Road
Summit, NJ 07901
908.277.1143 (W)
908.654.8972 (Fax)
6182700@mcimail.com (E-mail)

Clair Pollard
Math Specialist
Providence School Department
379 Washington Street
Providence, RI 02903
401.456.9476 (W)
401.456.9477 (Fax)
oats062@ri.ride.com. (E-mail)

Sandra Powers
Office of Undergraduate Studies
College of Charleston
Charleston, SC 29424-0001
843.953.5674 (W)
843.953.5544 (Fax)
powerss@cofc.edu (E-mail)

Nancy Puhlmann
Mathematics Teacher
Logan High School
Logan City School District
162 W. 100 S.
Logan, UT 84321
801.755.2380 (W)
801.755.2387 (Fax)
nancyp@1hs.logan.k12.ut.us (E-mail)

Caran Resciniti
Mathematics Department Chair
Fresno Systemic Initiative
1839 Echo Avenue
Fresno, CA 93204
209.441.3803 (W)

Diane Resek
Professor of Mathematics
San Francisco State University
8 Poppy Lane
Berkeley, CA 94708
415.338.2071 (W)
510.524.7329 (Fax)
resek@math.sfsu.edu (E-mail)

Barbara Reys
Department of Curriculum and Instruction
University of Missouri - Columbia
Columbia, MO 65211
573.882.8744 (W)
573.882.4481 (Fax)
cibr@showme.missouri.edu (E-mail)

Carol Rezba
Principal Mathematics Specialist
Virginia Department of Education
James Monroe Building
101 North 14th Street
Richmond, VA 23219-3684
804.225.2676 (W)
804.785.1703 (Fax)
crezba@pen.k12.va.us (E-mail)

Julia Robbins
Instructional Specialist
Rock Hill School District 3
District Office
P.O. Drawer 10072
Rock Hill, SC 29731
803.981.1055 (W)
803.981.1094 (Fax)
jrobbins@rock-hill.k12.sc.us (E-mail)

Mark Roddy
Assistant Professor
Seattle University
Broadway and Madison
Seattle, WA 98122-4460
206.296.5765 (W)
206.296.2053 (Fax)
mroddy@seattleu.edu (E-mail)

Mary Ann Rogers
Mathematics Teacher
J.H. Johnson Junior High School
DC Public Schools
1400 Bruce Place, S.E.
Washington, DC 20020
202.645.3700 (W)

Judy Roitman
Department of Mathematics
University of Kansas
Lawrence, KS 66045
785.864.4630 (W)
roitman@math.ukans.edu (E-mail)

Linda Rosen
Executive Director
National Council of Teachers of Mathematics
1906 Association Drive
Reston, VA 20191

Ilga Ross
Portland Community College
P.O. Box 19000
Portland, OR 97280-0990
503.977.4171 (W)
iross@pcc.edu (E-mail)

Mary Russick
Teacher, South Side Middle School
Pinellas County Schools
1701 10th Street South
St. Petersburg, FL 33705
813.893.2742 (W)
813.535.5180 (Fax)

Jim Sandefur
Department of Mathematics
Georgetown University
Washington, DC 20057
202.687.6145 (W)
202.687.6067 (Fax)
sandefur@math.acc.georgetown.edu (E-mail)

Diane Schaefer
Mathematics Specialist
Rhode Island State Department of Education
255 Westminster Street
Providence, RI 02903
401.277.4600 (W)
401.277.6033 (Fax)
ride0008@ride.ri.net (E-mail)

Donald Scheuer
Mathematics Department Chairman
Abington Junior High School
Abington School District
2056 Susquehanna Street
Abington, PA 19001
215.884.4700 (W)
donalds290@aol.com (E-mail)

Alan Schoenfeld
Education, EMST
University of California at Berkeley
Tolman Hall #1670
Berkeley, CA 94720-1670
510.642.0968 (W)
alans@socrates.berkeley.edu (E-mail)

Connie Schrock
Associate Professor of
Mathematics and Computer Science
Division of Mathematics and Computer Science
Emporia State University
Box 4027
Emporia, KS 66801
316.341.5631 (W)
316.341.6055 (Fax)
schrockc@esumail.emporia.edu (E-mail)

James Schultz
Morton Professor of Mathematics Education
Ohio University
117 McCracken Hall
Athens, OH 45701
614.593.4430 (W)
614.593.0477 (Fax)
schultzj@ouvaxa.cats.ohiou.edu (E-mail)

Bill Scott
Secondary Mathematics Supervisor
State Department of Public Instruction
301 N. Wilmington Street
Raleigh, NC 27601
919.715.1791 (W)
919.715.0517 (Fax)

Linda Selwood
K-12 Mathematics Coordinator
Concord High School
Brandywine School District
2501 Ebright Road
Wilmington, DE 19810
302.529.3082 (W)
302.529.3094 (Fax)
Selwood@copland.udel.edu (E-mail)

Vicki Shirley
Mathematics Teacher
Corinth High School
1310 Harper Road
Corinth, MS 38834
601.286.1000 (W)
601.286.1003 (Fax)
vfshirley@tsixroads.com (E-mail)

Todd Shuler
Texas Instruments
7625 Campbell Creek Lane
Charlotte, NC 28212
704.567.6122 (W)
704.567.2096 (Fax)

Judith Silbaugh
Mathematics Supervisor
Columbus Public Schools
Shepard Center
873 Walcutt Avenue
Columbus, OH 43219
614.365.5296 (W)
614.365.6260 (Fax)
jsilbaug@iwaynet.net (E-mail)

Ed Silver
LRDC/University of Pittsburgh
3939 O'Hara Street
Pittsburgh, PA 15260
412.624.3231 (W)
412.624.9149 (Fax)
eas@vms.cis.pitt.edu (E-mail)

Terry Souhrada
Professor, Department of Mathematical Sciences
The University of Montana
Missoula, MT 59801
406.243.2696 (W)
406.243.2674 (Fax)
terryas@selway.umt.edu (E-mail)

Carolyn Steele-Graham
K-12 Coordinator
Mathematics, Science and Technology
Troy City School
1950 Burdett Avenue
Troy, NY 12180
518.271.5381 (W)
518.274.2341 (Fax)

Hannah Slovin
Educational Associate, Mathematics Education,
Curriculum Research, and Development Group
University of Hawaii
1776 University Avenue
Honolulu, HI 96822

Myriam Steinback
Mathematics Specialist
Eisenhower Regional Alliance
for Mathematics and
Science Education Reform
TERC
2067 Massachusetts Avenue
Cambridge, MA 02140

Alberta Sullivan
Teacher, Memphis Systemic Initiative
3950 Summer Avenue
Memphis, TN 38122
901.325.5952 (W)
901.722.4557 (Fax)

Marcia Sward
Executive Director
Mathematical Association of America
1529 18th Street, N.W.
Washington, DC 20036
202.387.5200 (W)
202.387.5948 (Fax)
msward@maa.org (E-mail)

Mary Thompson
Mathematics Instructional Specialist
New Orleans Systemic Initiative
3510 General DeGaulle Drive
New Orleans, LA 70114
504.365.5608 (W)
504.365.5599 (Fax)

Cornelia Tierney
Senior Scientist
TERC
2067 Massachusetts Avenue
Cambridge, MA 02140
617.547.0430 (W)
617.349.3535 (Fax)
Cornelia_Tierney@terc.edu (E-mail)

Lisa Usher
Resource Teacher
Los Angeles Systemic Initiative
450 N. Grand Avenue, Room A-319
Los Angeles, CA 90012
213.625.6421 (W)
213.626.7785 (Fax)

Karen Usiskin
Senior Editorial Consultant
Scott Foresman - Addison Wesley
1900 E. Lake Avenue
Glenview, IL 60025
847.486.2415 (W)
847.486.3958 (Fax)
karen.usiskin@aw.com (E-mail)

Zalman Usiskin
Professor of Education
University of Chicago
5835 S. Kimbark
Chicago, IL 60637
773.702.1560 (W)
773.702.0248 (Fax)
z_usiskin@uchicago.edu (E-mail)

Irvin Vance
Professor of Mathematics
Department of Mathematics
Michigan State University
East Lansing, MI 48824
517.533.4693 (W)
517.432.1562 (Fax)
vance@math.msu.edu. (E-mail)

Martin van Reeuwijk
Freudenthal Institute
Tiberdreef 4
Utrecht University
3561 GG Utrecht, The Netherlands
31.30.261.1611 (W)
31.30.266.0430 (Fax)
martinr@fi.ruu.nl (E-mail)

Bert K. Waits
Professor Emeritus of Mathematics
Department of Mathematics
Ohio State University
231 W. 18th Avenue
Columbus, OH 43210
614.292.1934 (W)
614.292.0694 (Fax)
waits@math.ohio-state.edu. (E-mail)

Joan Waldvogel
Instructor, Developmental Mathematics
Portland Community College
P.O. Box 19000
Portland, OR 97280
503.978.5277 (W)
503.978.5050 (Fax)
jwaldvog@pcc.edu (E-mail)

Joy Wallace
Program Development
Science and Mathematics Consortium
for North West Schools
Columbia Education Center
171 NE 102nd Avenue
Portland, OR 97220
503.252.4999 (W)
joy@col_ed.org (E-mail)

E. Jean Ware
Curriculum Supervisor, 6-12 Mathematics
Caddo Parish
Caddo Parish School Board-Instructional Center
1961 Midway Avenue
Shreveport, LA 71108
318.638.6257 (W)
318.631.5241 (Fax)

Jesse Warren
Curriculum Facilitator
South Bend Community School Corporation
635 S. Main Street
South Bend, IN 46601
219.283.8057 (W)
219.283.8059 (Fax)
jwarren@sbcsc.k12.in.us (E-mail)

Stacey Weinand
Interim State Mathematics Coordinator
4712 Tanglewood Court
Norman, OK 73072
405.447.4103 (W)
405.521.6205 (Fax)
dweinand@prodigy.net (E-mail)

Marvin Weingarden
Mathematics Supervisor
Detroit Systemic Initiative
Detroit Public Schools, Room 932
5057 Woodward Avenue
Detroit, MI 48202
313.494.1608 (W)
313.494.7864 (Fax)

Cathy Wick
Assistant Professor
Department of Mathematics, ECC139
St. Cloud State University
720 4th Avenue, South
St. Cloud, MN 56301
320.255.3001 (W)
320.255.4262 (Fax)
cathy@tigger.stcloud.msus.edu (E-mail)

Ted Widerski
K-12 Mathematics Resource Teacher
Madison Metropolitan School District
545 W. Dayton Street
Madison, WI 53703
608.266.6158 (W)
608.261.9022 (Fax)
widerski@smallbytes.net (E-mail)

Lorna Wiggan
Director, National Council
of Teachers of Mathematics
Toronto Board of Education
165 Dixon Road, Unit 57
Etobicoke, Ontario M9P 3T9 Canada
416.397.3787 (W)
lwiggan@interlog.com (E-mail)

Sharon Wiggs
Curriculum Specialist
Baltimore City Schools
200 East North Avenue
Baltimore, MD 21202
410.396.8585 (W)
410.396.8063 (Fax)
shwiggs@aol.com (E-mail)

Beverly Williams
K-12 Mathematics Curriculum Coordinator
Pulaski County Special School District
925 East Dixon
P.O. Box 8601
Little Rock, AR 72216
501.490.2000 (W)
501.490.0254 (Fax)
bwilliam@pcssd1.k12.ar.us (E-mail)

Susan Williams
Assistant Professor
College of Education
Curriculum and Instruction Department
University of Houston
Houston, TX 77204-5872
713.743.4969 (W)
713.743.4990 (Fax)
sewilliams@uh.edu (E-mail)

Virginia Williams
Manager, Member Relations
National Council of Teachers of Mathematics
1906 Association Drive
Reston, VA 20191
703.620.9840 (W)
703.476.2970 (Fax)

Ruth Wilson
Secondary Mathematics Specialist
Clark County School District
601 North 9th Street
Las Vegas, NV 89107
702.799.8436 (W)
702.362.3059 (Fax)
wilson@fremont.vegas.net (E-mail)

Susan Wood
Professor, J. Sargeant Reynolds Community College
10402 Warren Road
Glen Allen, VA 23060-3039
804.672.6527 (W)
aewood@erols.com (E-mail)

Rich Wyllie
Mathematics Instructor
Community High School District 99
1436 Norfolk
Downers Grove, IL 60516
630.271.6611 (W)
630.968.3740 (Fax)
kingcalc@aol.com (E-mail)

Linda Young
Indiana Council President
Mathematics Teacher
Columbus East High School
Bartholomew Consolidated School Corp.
230 S. Marr Road
Columbus, IN 47201
812.376.4348 (W)
812.376.4358 (Fax)
eastmath@hsonline.net (E-mail)

Gisele Zangari
ATMIM President and
Massachusetts Statewide
Mathematics Coordinator
Mansfield Public School and the Massachusetts
Department of Education
Palms MA Department of Education
350 Main Street
Malden, MA 02148
617.388.3300 (W)
gzangari@doe.mass.edu (E-mail)

Rose Mary Zbiek
Assistant Professor of Mathematics Education
and Mathematics
University of Iowa
289 Lindquist Center North
Iowa City, IA 52242-1529
319.335.6439 (W)
319.335.5608 (Fax)
rose-zbiek@uiowa.edu (E-mail)

Janie Zimmer
Mathematics Supervisor
Howard County Public Schools
10910 Route 108
Ellicott City, MD 21042-6198
410.313.6615 (W)
410.313.6835 (Fax)
jzimmer@co.howard.k12.md.us

Linda Zimmerman
Mathematics Test Specialist
ACT Inc.
2201 North Dodge Street
P.O. Box 168
Iowa City, IA 52243-0168
319.337.1741 (W)
319.339.3021 (Fax)
zimmer1@act.org (E-mail)

Appendix C

Speaker Biographies

Geoffrey Akst serves as Professor of Mathematics at the Borough of Manhattan Community College/The City University of New York, where he coordinates the college's developmental mathematics program. For more years than he would care to remember, he has taught algebra with varying combinations of the three C's—calculators, computers, and calcite (chalk!)—and has been the principal investigator for several grants dealing with the impact of technology on the college mathematics curriculum. Dr. Akst is currently preparing a manuscript on the changing nature of elementary algebra across the last 200 years as seen through the prism of English-language textbooks.

Hyman Bass received his PhD in Mathematics from the University of Chicago. In 1975, after 16 years in teaching and research positions, he accepted the position of Chairman of the Mathematics Department at Columbia University, where he remains as professor. As an academic mathematician, Dr. Bass is involved in basic research, mainly in algebra and its interface with geometry, and in teaching at all university levels. He is an author of research and expository writings, active in the editorial aspects of scientific publishing, and engaged in matters of educational and scientific policy. Dr. Bass has served as a member of or consultant to many national and international groups concerned with mathematical research and science. He has served on the Executive Committee for the American Mathematical Society and as Chairman of the Board of Trustees for the Mathematical Sciences Research Institute in Berkeley. Dr. Bass is a member of the National Academy of Sciences. He chaired an Advisory Committee on Pure Mathematics in 1970 for the National Research Council (NRC) and currently chairs the Mathematical Sciences Education Board (MSEB) for the NRC.

Rick Billstein is a Professor of Mathematics at the University of Montana in Missoula, MT. He has worked in the area of mathematics education for 30 years and is probably best known as a co-author of the book, *Mathematics for Elementary Teachers: A Problem Solving Approach,* which is now in its 6th edition. He is presently the Director of the National Science Foundation-funded "Six Through Eight Mathematics" (STEM) project. STEM has a five-year grant to develop a new standards-based mathematics curriculum for middle schools. The materials are now available through McDougal Littell/Houghton Mifflin.

Jere Confrey joined the faculty of the University of Texas at Austin as a Professor of Mathematics Education in 1997. Her research at Cornell University has focused on how to reform the teaching of algebra toward the ideas of functions and modeling. She works on these topics across the curriculum, starting with young children's understanding of multiplication, division, and ratio and proceeding through the grades toward the introduction of calculus. She is the author of the software, "Function Problem," and co-author of "Multimedia Precalculus" (in progress). She was the founder of the SummerMath program for young women at Mount Holyoke College and co-founder of SummerMath for Teachers. She believes that research is best developed in relation to ongoing classroom practices and, thus, typically is involved in extensive school-based activities. Her most recent work has been with seventh grade students and teachers in designing and implementing a technology-rich introduction to algebra. She is a senior editor of the *International Journal on Computer and Mathematics Learning* and is a member of the editorial panel of the *Journal of Research in Mathematics Education.*

Rosalie Dance teaches mathematics at Georgetown University, where she also develops mathematical modeling activities for use in secondary and community college classrooms. Her research interest centers on means to promote equity in mathematics education, and she works as a consultant on mathematics education and equity issues. She was a District of Columbia Public Schools mathematics teacher for an incredibly long time and received a Presidential Award for Excellence in Mathematics Teaching. She has led many inservice courses and workshops for secondary mathematics teachers. She has been a mathematics teacher and a mathematics student in six countries on three continents.

Robert L. Devaney is currently Professor of Mathematics at Boston University. His main area of research is dynamical systems theory. He is currently director of the Dynamical Systems and Technology Project at Boston University. This is a National Science Foundation-sponsored project to show teachers how to incorporate modern ideas in mathematics into the high-school curriculum, as well as to provide teachers with the necessary technological tools.

John A. Dossey, Distinguished University Professor of Mathematics at Illinois State University, served as President of National Council of Teachers of Mathematics from 1986-1988 during the writing of the *Curriculum and Evaluation Standards for School Mathematics.* Dr. Dossey is currently the Chair of the Conference Board of Mathematical Science and chaired the commission appointed to develop the dimensions of and specifications for the proposed national mathematics exam.

Gregory Foley is a mathematics educator, consultant, and textbook author and is interested in the appropriate uses of technology for mathematics teaching and learning in grades 6-14. Dr. Foley obtained BA and MA degrees in mathematics and a PhD in mathematics education from the University of Texas at Austin. He is an associate professor of mathematics at Sam Houston State University in Huntsville, TX, having previously held faculty positions at North Harris County College, Austin Community College, and Ohio Sate University. Dr. Foley has presented some 120 papers, talks, lecture-demonstrations, workshops, minicourses, and summer institutes in 28 states, Belgium, Canada, England, Mexico, and Spain. He has directed three federally supported calculus projects and contributed his services to the American Mathematical Association of Two-Year Colleges, the Mathematical Association of America, the Mathematical Sciences Education Board, and CBAMN.

Landy Godbold currently holds the Loridans Foundation Alex P. Gaines Chair and teaches mathematics at The Westminster Schools. He has taught at the Westminster Schools since 1975. He received both his BS and MS degrees in Applied Mathematics from Georgia Institute of Technology and his EdS in Secondary Mathematics Curriculum and Instruction from Georgia State University. He served for six years as Computer Director for the Woodrow Wilson National Fellowship Foundation Summer Institutes. He is a member of the Mathematical Association of America and the National Council of Teachers of Mathematics, he has contributed to the *Mathematics Teacher,* and he has spoken at numerous regional and national conferences. He is currently Principal Investigator for ARISE, one of several National Science Foundation-funded secondary curriculum projects.

Deborah Hughes Hallet is a Professor of the Practice in the Teaching of Mathematics at Harvard University. She has also taught at the Middle East Technical University in Ankara, Turkey, and at the University of Arizona. She graduated from Cambridge University in England. Professor Hughes Hallet served as Principal Investigator, with Andrew Gleason and David Mumford, for the National Science Foundation-funded "Calculus Consortium," which is based at Harvard.

Christian Hirsch is a Professor of Mathematics and Mathematics Education at Western Michigan University, Kalamazoo, MI. He received his PhD in mathematics (with specialization in mathematics education) from the University of Iowa in 1972. Dr. Hirsch was a member of the National Council of Teachers of Mathematics'

(NCTM) Commission on Standards for School Mathematics and Chair of its Working Group on Curriculum for Grades 9-12. He is co-author of several high-school mathematics textbooks, author of numerous articles in mathematics education journals, and editor of several NCTM publications. Dr. Hirsch has served as President of the Michigan Council of Teachers of Mathematics, on the Board of Directors of the NCTM, and on the Board of Directors of the School Science and Mathematics Association. Dr. Hirsch currently directs the "Core-Plus Mathematics Project," a National Science Foundation-funded, comprehensive high-school curriculum development project. In addition, for the past five years he has directed the "Making Mathematics Accessible to All" project, an Eisenhower-funded statewide professional development program for teachers.

James J. Kaput, with a PhD in Mathematics from Clark University, joined the faculty at what eventually became the University of Massachusetts at Dartmouth in 1968, where he is now Chancellor Professor of Mathematics. He has written or edited a half dozen books and is currently finishing an edited volume summarizing work done by the Algebra Working Group of the National Center for Research in Mathematical Sciences Education from 1989-1995, which involved rethinking the role of algebra in school mathematics and researching the early development of algebraic reasoning in the context of elementary mathematics. Dr. Kaput's current National Science Foundation-funded "SimCalc Project" involves building and testing simulations and visualization tools on both computers and calculators for learning the fundamental ideas underlying calculus beginning at the elementary-school level. Dr. Kaput is on the editorial board of five journals related to mathematics education and is a founding co-editor, with Alan Schoenfeld and Ed Dubinsky, of a new series of volumes jointly published by the American Mathematical Society and the Mathematical Association of America on research in collegiate mathematics education. He is on many research and development project advisory boards, publishes widely, and is a frequent speaker at national and international meetings

Glenn Kleiman is Vice President and Senior Scientist at the Education Development Center, Inc. (EDC) and lecturer in the Technology in Education program at the Harvard Graduate School of Education. A cognitive psychologist by background (PhD, Stanford University, 1977), Dr. Kleiman has long been involved in research on children's learning, exploring applications of technology in education, and developing innovative curriculum materials and educational software. He has directed two large-scale mathematics curriculum development projects funded by the National Science Foundation, the most recent of which developed the "Mathscape: Seeing and Thinking Mathematically" middle-school curriculum, published by Creative Publications.

Glenda Lappan, Department of Mathematics, Michigan State University (MSU), received her EdD in Mathematics and Education from the University of Georgia in 1965. She has been a member of the Department of Mathematics faculty at MSU since she received her degree. Her research and development interests are in the connected areas of students' learning of mathematics and the professional development of mathematics teachers at the middle and secondary levels. She is currently Co-Director of the "Connected Mathematics Project," President-Elect of the National Council of Teachers of Mathematics (NCTM), Vice-Chair of the Mathematical Sciences Education Board, and a member of the National Education Board for Research Policy and Priorities. She served as Chair of the Middle-School Working Group for the NCTM *Curriculum and Evaluation Standards for School Mathematics* and was Chair of the commission that developed the NCTM *Professional Standards for Teaching Mathematics.*

David D. Molina, PhD, serves as the Associate Director of the Charles A. Dana Center and as Adjunct Professor of Mathematics Education at the University of Texas at Austin. A former high-school mathematics teacher and experienced higher education faculty member, Dr. Molina has made significant contributions to textbook publishing, professional development, teacher preparation, and educational technology. His current work spans the areas of mathematics education policy, curriculum development, algebra reform, systemic change, teacher preparation reform, and school restructuring.

Beatrice Moore-Harris is currently a K-12 Mathematics Instructional Specialist for the Houston Independent School District's South Central District. She has worked as an Educational Specialist (Mathematics and Title I) for the Region IV and Mathematics Specialist, K-8, in Fort Worth Independent School District. The majority of her work as teacher and supervisor has been with urban school districts that are addressing the challenge of preparing all students to successfully complete Algebra I, increasing the postsecondary attendance rate of underrepresented youth, and coordinating professional development and student academic enrichment activities. Ms. Moore-Harris has served as Chair of the *Mathematics Teaching in the Middle School* journal's Editorial Panel and Editor of the "Mathematics Teaching in the Middle School Algebra Focus Issue." She was a member of the Algebra Working group, which produced the document entitled "Creating a Vision for Algebra." She is co-author of *Mathematics Applications and Connections, Courses 1, 2, 3,* and *Algebra I and Algebra II.* She is consulting author for *Interactive Mathematics, Units 1-18,* published by Glencoe.

Milton O. Norman is a member of the mathematics department at Granby High School in Norfolk, VA. He received his BS degree from Old Dominion University in Norfolk and his MBA in Information Systems from Golden Gate University in San Francisco. Mr. Norman retired from military service to begin teaching at Granby High School. He is constantly searching for an effective means of mentoring students to help them appreciate the beauty of mathematics. He has served as a co-facilitator of a series of workshops for mathematics and science teachers entitled "Enhancing Instruction through Technology." He is an instructor for the Virginia Network for Technology. Since 1990, Mr. Norman has been "Teacher of the Year" at Granby High School twice, and since 1990, he has received the "School Bell Award" from Norfolk Public Schools for outstanding contributions to education six times. He is a national instructor with "Teachers Teaching with Technology."

Elizabeth Phillips is a member of the Department of Mathematics at Michigan State University. She has been the director or principal investigator of several teacher enhancement, curriculum development, and research grants in mathematics education at both the middle- and high-school levels. She was Editor of the Michigan Council of Teacher of Mathematics Monograph series. In 1990, she served as a member of the "Leading Mathematics into the 21st Century" project. She is author of the National Council of Teachers of Mathematics (NCTM) Addenda book, *Patterns and Functions for Middle Grades Teachers* (NCTM, 1991). Ms. Phillips conducts numerous in-service workshops for teachers, gives speeches, and is the author of several papers and books, including the *Middle Grades Mathematics Project.* She was Co-Chair of the National Science Foundation (NSF)-funded Algebra for the Twenty-First Century conference and Chair of the NCTM Algebra Working Group. Currently, she is Co-Director of "The Connected Mathematics Project," a five-year NSF-funded project to write, test, and implement a complete mathematics curriculum for the middle grades. She is also Co-Director of an NSF-funded teacher enhancement grant that ends in 1999. These projects complement her ongoing interest in the learning and teaching of mathematics.

Henry Pollak spent 35 years at Bell Telephone Laboratories and Bell Communications Research doing mathematical research and heading various research organizations devoted to this and related topics. He received his BA from Yale University and his MA and PhD from Harvard University. In addition to his research in various areas of mathematics related to communications, Dr. Pollak has been active in mathematics education. He has served as a member of the Mathematical Sciences Education Board, Chairman of the School Mathematics Study Group Advisory Board, Chairman of the Advisory Committee to the National Science Foundation's Education Directorate, President of the Mathematical Association of America, member of the Board of Trustees for the North Carolina School for Science and Mathematics, member of the COMAP Board of Trustees, and Co-Director of the Woodrow Wilson National Fellowship Foundation's summer institutes for mathematics teachers. Dr. Pollak retired from Bell Communications Research in 1986 and is now a Visiting Professor at Teachers College of Columbia University.

Diane Resek is a Professor of Mathematics at San Francisco State University and one of the authors of the National Science Foundation-funded secondary curriculum project, "The Interactive Mathematics Program." She has taught and written curriculum materials for all levels of mathematics, from pre-school to graduate school.

Cornelia Tierney is a researcher with Ricardo Nemirovsky on children's understanding of the mathematics of change and an author of the National Science Foundation-sponsored K-5 curriculum, "Investigations in Number, Data, and Space." A coordination of work on these two projects resulted in the writing of curriculum units for grades 2-5 that involve children in investigating and representing changing situations. Dr. Tierney was for many years a classroom teacher at the middle school level. She is interested in how the mathematics of change contributes to the development of ideas basic to the study of algebra.

Martin van Reeuwijk is a mathematics educator and researcher. He is the Dutch coordinator of the "Mathematics in Context" project and one of the authors in the project. His interests include assessment, algebra (school algebra), and the use of new technologies in mathematics education.

Bert K. Waits is Professor Emeritus of Mathematics and visiting Professor of Mathematics, Science, and Technology Education at Ohio State University. He is the co-founder of the international T-cubed (Teachers Teaching with Technology) professional in-service teacher development program that was founded at Ohio State in 1987. In 1987 and 1988, Dr. Waits served as a member of the National Council of Teachers of Mathematics Working Group that wrote the *Curriculum and Evaluation Standards for School Mathematics* for grades 9-12. His current focus is mathematics teacher-training and the appropriate use of hand-held technology in the teaching and learning of mathematics. Since 1988, Dr. Waits has assisted in the design of graphing calculators and hand-held computers.

Beverly Williams is the K-12 Mathematics Curriculum Coordinator for the Pulaski County Special School District in Arkansas (near Little Rock). Her interest in the teaching and the learning of algebra is an ongoing theme in her professional activities. In 1989, she received a Woodrow Wilson National Fellowship Foundation (WWNFF) award to Princeton University for the study of algebra. Additionally, since 1990 she has been a TORCH teacher on a WWNFF Algebra Team, doing workshops around the nation on algebra: its curriculum, the impact of technology, and the role of multiple representations. She is a member of the Advisory Committee for the PBS Math Line Algebra Project. In addition to her algebra affiliations, she brings to this algebra symposium her experiences as a member of the National Council of Teachers of Mathematics Conventions and Conference Committee, from 1993-1996, and, from 1996-1997 as chair of the committee.

Rose Zbiek currently teaches mathematics and mathematics education courses for undergraduate and graduate students at the University of Iowa. She previously taught mathematics and computer science courses at Lake-Lehman High School in Northeastern Pennsylvania. Her interests center on using computing tools to develop mathematical understanding, with an emphasis on the roles of justification, proof, and evidence. She is a co-author of *Computer-Intensive Algebra, Concepts in Algebra: A Technological Approach* and the addenda booklet, "Algebra in a Technological World."

Appendix D

Invited Papers

Stating the Obvious: Mathematics Course Taking Matters (W. Tate)

Algebra, Technology, and a Remark of I.M. Geland (M. Saul)

Stating the Obvious: Mathematics Course Taking Matters

William F. Tate
University of Wisconsin at Madison
Madison, Wisconsin

Many studies indicate that a disproportionate amount of minority and low SES students receive a mathematics education more closely associated with basic skills (Oakes, 1990; Secada, 1992). In the late 1960s and early 1970s, a mathematics reform movement, "back to basics," emerged, which focused largely on elementary and middle schools (NCTM, 1980). This movement was partly a product of efforts to achieve equality of educational opportunity through compensatory education. The back-to-basics effort called for teaching a core set of rudimentary mathematics procedures and facts, often to the exclusion of more advanced mathematical ideas. Although the basic skills movement influenced the entire educational system in the United States, it had a particularly significant impact on the mathematics curriculum and pedagogy in low-income urban and rural schools (Strickland and Ascher, 1992). On the positive side, the basic skills movement did result in improved standardized mathematics test scores for students who were traditionally underserved (Secada, 1992). It illustrated that when teachers and principals agreed on a common standard in mathematics and received adequate institutional support to achieve the standard, students would learn the content.

Another reason for the success of the basic skills movement had to do with teachers' beliefs. Much of the basic skills philosophy was consistent with teachers' perceptions of student ability (Zeichner, 1996; Knapp and Woolverton, 1995). However, as the vision of what it means to know and understand mathematics has shifted from the basic skills standard to a more demanding goal, the limits of past pedagogical practice have become obvious. Teaching strictly lower level, basic math skills to students throughout their elementary- and middle-school years will not adequately prepare them for the challenges of college preparatory mathematics. Moreover, the practice of tracking students out of college prep mathematics opportunities is akin to restricting them to the basic skills curriculum.

SOME RECENT TRENDS: COURSE TAKING MATTERS

The impact of restricting students to a basic skills mathematics curriculum is most apparent when studies of course taking and mathematics achievement are examined. Specifically, if the goal is high standards for all students, then a close examination of the influence of course-taking effects is essential.

Hoffer and colleagues (1995) reported on the relationship between the number of mathematics courses that high-school students of different racial and SES backgrounds completed and their achievement gain from the end of grade 8 to grade 12. Hoffer et al. indicated that when African American and white students who completed the same courses were compared, the difference in average achievement gains was smaller than in other circumstances, and none was statistically significant. Similarly, Asian and white students' mathematics achievement gains were smaller and also generally reduced among students completing the same number of mathematics courses. Moreover, none of the SES comparisons showed significant differences among students taking the same number of courses. These findings can be interpreted to mean that much of the racial and SES differences in mathematics achievement in grades 9-12 is the product of the different numbers of mathematics courses that African American and white, Asian and white, and high and low SES students complete during secondary school.

Smith (1996) investigated the efforts of early access (eighth grade) to algebra on students' access to advanced mathematics courses and subsequent high-school mathematics achievement. She found that early access to algebra has an effect beyond increased achievement and, in fact, may socialize a student into taking more mathematics. In essence, having credit for a year of algebra at the beginning of high school is a credential, regulating access to more advanced course work in mathematics. Moreover, having this credential increases both the students' and educators' expectations about how much mathematics the student will take in high school, keeping students in the college prep track longer and producing higher achievement as a result.

The importance of these two studies is that policy can be used to intervene.[1] One method of intervention is to mandate specific course requirements at the eighth grade and/or secondary level. To enhance policy intervention, student learning is required at the elementary level. Clearly, mandating more secondary courses without systemic efforts to change elementary-school mathematics experiences and achievement levels is potentially problematic. Further, mandates come with costs—avoidance and implementation costs.

FINAL REMARKS: IMPLEMENTATION AND AVOIDANCE COSTS

Like all policies, mandating mathematics courses has costs associated with it. Space limitations prevent detailed discussion of those costs. However, it is important to note that school systems must examine their practices and goals closely to determine both the social and fiscal costs related to mandating and implementing courses. Many schools have mandated "algebra for all," then constructed multiple tiers of algebra courses. Thus, the policy is implemented on paper, but the intent of the policy is compromised. The costs associated with this kind of practice are avoidance costs. Another tactic that might be employed is to mandate eighth grade "algebra for all," then to create a separate track for a select group of students to take algebra in seventh grade or earlier, thus recreating the credentialing system described by Smith (1996).

I charge teachers and administrators contemplating the mandating of courses to examine implementation and avoidance costs connected to this important policy matter. The future of many underserved students is at stake.

REFERENCES

Hoffer, T. B., Rasinski, K.A., and Moore, W. (1995.) *Social Background Differences in High School Mathematics and Science Course Taking and Achievement* (NCES 95-206). Washington, DC: U.S. Department of Education.

Knapp, M.S., and Woolverton, S. (1995.) "Social Class and Schooling." In J.A. Banks and C. Banks (eds.), *Handbook of Research on Multicultural Education* (pp. 548-569). New York: Macmillan.

National Council of Teachers of Mathematics (NCTM). (1980.) *An Agenda for Action: Recommendations for School Mathematics of the 1980s*. Reston, VA: Author.

Oakes, J. (1990.) "Opportunities, Achievement, and Choice: Women and Minority Students in Science and Mathematics." In C.B. Cazden (ed.), *Review of Research in Education* (pp. 153-222). Washington, DC: American Educational Research Association.

Secada, W.G. (1992.) "Race, Ethnicity, Social Class, Language, and Achievement in Mathematics." In D. A. Grouws (ed.), *Handbook of Research on Mathematics Teaching and Learning* (pp. 623-660). New York: Macmillan.

Smith, J.B. (1996.) "Does an Extra Year Make Any Difference? The Impact of Early Access to Algebra on Long-Term Gains in Mathematics Achievement." *Educational Evaluation and Policy Analysis*, 18(2), 141-153.

Strickland, D. S., and Ascher, C. (1992.) "Low-Income African American Children and Public Schooling." In P.W. Jackson (ed.), *Handbook of Research on Curriculum* (pp. 609-625). New York: Macmillan.

Zeichner, K. M. (1996.) "Educating Teachers to Close the Achievement Gap: Issues of Pedagogy, Knowledge, and Teacher Preparation." In B. Williams (ed.), *Closing the Achievement Gap: A Vision for Changing Beliefs and Practices* (pp. 56-76). Alexandria, VA: Association for Supervision and Curriculum Development.

[1]In contrast, it is nearly impossible to intervene in or change teacher beliefs. If teachers' beliefs are inconsistent with the policy proposed, a significant hurdle is created for the policy-maker interested in implementing a reform—e.g., algebra for all.

Algebra, Technology, and a Remark of I.M. Gelfand

Mark Saul
Bronxville School
Bronxville, New York

SOME VIEWS OF ALGEBRAIC THINKING

Historically, the development of algebra grew from a consideration of the arithmetic of the rational numbers[1] and particularly from the investigation of methods of solution of polynomial equations. The introduction of variables to make general statements about rational numbers, or to pose and solve problems involving them, was a natural step, taken very early.

The conception of the nature of algebra behind these efforts is perhaps most easily seen in Isaac Newton's book on elementary algebra, which bore the title *The Universal Arithmetic.*[2] Newton himself gives a clear exposition of the philosophy:

> "Common arithmetic and algebra rest on the same computational foundations and are directed to the same end. But whereas arithmetic treats questions in a definite, particular way, algebra does so in an indefinite, universal manner, with the result that almost all pronouncements which are made in this style of computation—and its conclusions especially—may be called theorems. However, algebra most excels, in contrast with arithmetic where questions are solved merely by progressing from given quantities to those sought, in that for the most part it regresses from the sought quantities, treated as given, to those given, as though they were the ones sought, so as...to attain some conclusion—that is—equation—from which it is permissible to derive the quantities sought... . Yet arithmetic is so instrumental to algebra in all its operations that they seem jointly to constitute but a unique, complete computing science, and for that reason I shall explain both together."[3]

An essentially different view of algebra emerged in the twentieth century. Morris Kline points out its roots even earlier:

> "Algebra, for Descartes, precedes the other branches of mathematics... . There is a sketch of a treatise on algebra, known as Le Calcul (1638), written either by Descartes himself or under his direction, that treats algebra as a distinct science. His algebra is devoid of meaning. It is a technique of calculation, or a method, and is part of his general search for method... . Descartes' view of algebra as an extension of logic in treating quantity suggested to him that a broader science of algebra might be created, which would embrace other concepts than quantity and be used to approach all problems."[4]

---

[1]See, for example, Victor J. Katz, "The Development of Algebra and Algebra Education," in *The Algebra Initiative Colloquium*, edited by Carole B. Lacampagne, William Blair, and Jim Kaput. (Washington, DC: U.S. Department of Education, Office of Educational Research and Improvement, 1995), 15-30.

[2]D.T. Whiteside, editor. *The Mathematical Papers of Issac Newton,* Volume V, 1683-1684 (Cambridge: At the University Press, 1972), 538ff.

[3]*Ibid.*, 539.

[4]Morris Kline, *Mathematical Thought from Ancient to Modern Times* (New York: Oxford University Press, 1971), 281.

It was a bit early for this bold step, and Descartes did not complete this work. However, Kline traces echoes of its influence on the work of Leibniz and De Morgan (1187-1189) in developing mathematical logic.[5]

It was not until the twentieth century that the conception of algebra as the study of manipulation of certain symbols acquired meaning. Katz notes that "it was Emmy Noether who taught the mathematicians in Gottengen in the 1920s that algebra was central to mathematics, that its ideas extended to all areas of the subject, and that an abstract approach was the way to look at algebraic concepts." [6]

A REMARK OF I.M. GELFAND

I.M. Gelfand offers a twentieth century view of the nature of algebraic thinking that has meaning both for mathematics and for pedagogy.[7] His remark: "Arithmetic is essentially the study of the field of rational numbers. Algebra begins with the study of the field of rational expressions."

This remark draws an arbitrary—but clear—line between arithmetic and algebra. It highlights the fact that the essence of algebra is not simply the use of variables. It also distinguishes the study of algebra from the study of functions.

Gelfand illustrates his idea with two examples. A common activity for middle-school or upper elementary students is the "guess my number" trick. For example, the teacher asks the student to pick a number, double it, add three, then subtract the original number. If the student gives the answer, the teacher can easily "guess" the number that the student had originally chosen. The learning activity is then for the student to explain the trick, to be able to use it herself, or to make up an original "trick" of the same sort.

This activity is an excellent one to get students to generalize statements about arithmetic. In trying to express what is going on, they often are led directly to the use of variables. But there is another level to algebra, the level of computation with rational expressions, which it does not reach.

This level is reached when students learn patterns of factoring, for example. Seeing the factorization $A^2 - B^2 = (A + B)(A - B)$, a student who has mastered certain algebraic concepts can factor, $x^4 - y^4$, $4x^2 - 1$, or $\cos^2 x - \sin^2 x$. Mathematically, the student is then letting the variables A and B stand for other rational expressions (or rational trigonometric expressions). Crudely put, in arithmetic, letters are used to stand for numbers; in algebra they are used to stand for other letters.

Gelfand's statement is a type of definition, and, like all definitions, it is arbitrary. It is an attempt to use language to distinguish one set of mathematical concepts from others. We will make the argument that this particular distinction is useful and enlightening, both from a mathematical point of view and from a pedagogical perspective.

ALGEBRA AND PEDAGOGY

Gelfand's remark implies a view of algebra different from that espoused by several recent writers on learning and teaching. For example, Judah Schwartz has said, "A proposal is made to restructure the post-arithmetic mathematics curriculum around the function (and the entailed concept variable) as the central, and indeed the only necessary mathematical and pedagogical object of the subjects now called algebra, trigonometry, pre-calculus, and calculus."[8] This is a pedagogical remark, not a mathematical one, and perhaps Schwartz had in mind the courses labeled "algebra" in many schools and textbooks. A strong argument can be made for using the function concept as a unifying theme in the development of students' mathematical thought. However, we will make the argument that, even pedagogically, it is important to distinguish algebraic thinking from thinking about functions, regardless of the judgment that is then made about the role of either in education.

The conception of algebraic thinking implied by Gelfand's remark is closer to that implied by some notes by Kieran and Chalouh: "Most algebra courses begin immediately with the use of letters as mathematical objects and

[5]*Ibid.*, 1187-1189.

[6]*Ibid.*, 30.

[7]Personal communication to the author, November 1995.

[8]Quoted in *Algebra for the Twenty-First Century: Proceedings of the August 1992 Conference.* (Reston, VA: NCTM, 1992), 26.

then proceed to the operations that can be carried out on these objects."[9] Kieran and Chalouh go on to describe the experiences that students need to understand algebra: "Links between the use of numbers in arithmetic and the use of letters in algebra are rarely accorded more than a passing nod in high school algebra."[10] Implicit in these comments is the idea that algebra is different from its application to arithmetic.

We argue below that algebra is a separate entity from either the study of functions or a generalization of arithmetic but that both are important applications of the central ideas of algebra.

HOW IS THE STUDY OF ALGEBRA RELATED TO THE STUDY OF ARITHMETIC?

The fact that students can "learn" algebra without understanding its application as a generalization of arithmetic poses problems for every teacher of the subject. The following two vignettes reflect the author's own classroom practice in grappling with these problems.

Vignette 1

This scene took place some years ago in an urban high school. The teacher (the present writer) was working on a fairly standard New York State Regents curriculum with a group of well-motivated, high-ability ninth graders. The lesson took place in April, after the class had covered several topics in the factoring of polynomials.

The teacher wrote on the board:

$$\begin{array}{ll} 4 \times 4 = 16 & 3 \times 5 = 15 \\ 5 \times 5 = 25 & 4 \times 6 = 24 \\ 6 \times 6 = 36 & 5 \times 7 = 35 \\ 7 \times 7 = 49 & 6 \times 8 = 48 \\ \vdots & \vdots \\ 15 \times 15 = 225 & 14 \times 16 = 224 \end{array}$$

(This scene took place before calculators were readily available. A more modern practice might be to structure the examples so that the students, using the calculator, would get the answers to the multiplication problems themselves.)

The teacher then asked the class two questions:

1) Multiply 149×151 mentally.
2) Factor the number 3599.

The students in the class had learned all the "factoring patterns" necessary for success in the standard algebra course and, in particular, had some facility with the fact that $(x - 1)(x + 1) = x^2 - 1$. They could factor such expressions as $4a^2 - 1$, $9x^2 - 4y^2$, and even $(a + b)^2 - (a - b)^2$ with relative ease. Yet many of them were puzzled when faced with this situation.

Tom, for example, engaged in the following dialogue with the teacher:

Tom: But what are the dots for?
Teacher: Well, what comes after $7 \times 7 = 49$?
Tom: 8×8. [His voice indicates that he has come to the end of his response.]
Teacher: Then?
Tom: ...equals 64.
Teacher: Good. And then?
Tom: $9 \times 9 = 81$.
Teacher: Excellent. And what is in the right column? [He points to the space described.]

---

[9]Kiernan, Carolyn, and Louise Chalouh, "Prealgebra: The Transition from Arithmetic to Algebra" in *Research Ideas for the Classroom: Middle Grades Mathematics,* Douglas T. Owens, editor (Reston, VA: NCTM, 1993), 179.

[10]*Ibid.*

Tom: $8 \times 10 = 80$.
Teacher: Why?
Tom: Well, the number before 9 is 8, and the number after 9 is 10.
Teacher: OK. Write the next line.
Tom: [He fills in both columns of the next line, then makes a statement.] I see. The dots mean "and so on..." So... [He pauses.] 149×151 is the same as 150×150.
Teacher: The same?
Tom: Oh. It's one more. Less. [He says this in a loud voice, as if to erase the previous error.] [He was then able to solve both problems posed.]
Teacher: Now suppose we wrote $x \times x$ here [in the left column]. What could you write here [in the right column]?
Tom: $(x - 1)$ and $(x + 1)$.
Teacher: And what is $(x - 1)(x + 1)$?
Tom: $x^2 - 1$. Oh, cool. That says that the numbers here [right column] are one less than these numbers [left column]. I knew that.

Discussion of Vignette 1

Tom is physically a large young man who plays baseball with older students. At this point, his frame is trembling with excitement at the fact that he can understand a difficult problem. And the teacher also is excited both about Tom's experience and about the teacher's own discovery that Tom's thorough knowledge of factoring is different, on a cognitive level, from an understanding of the application of factoring to the description of numerical patterns.

The teacher's (my) experience with Tom is typical: students who "know" algebra—and even who test well on traditional algebraic subject matter—may not see how algebra can tell us something about the arithmetic of rational numbers.

Can we say that they do not understand algebra? They certainly understand how to work with rational expressions. What they do not yet see is the application of these expressions to the arithmetic of rational numbers.

Learning algebra may or may not involve learning a "universal arithmetic." As Kieran and Chalouh point out, the application of algebra to arithmetic requires a different set of experiences than the acquisition of purely algebraic skills.

Vignette 2

This scene took place in a small remedial class. The tenth grade students were well motivated but had problems learning the arithmetic of fractions and negative numbers.

Marion, for example, subtracts negative numbers by considering eight cases. If she is working on $a - b$, she considers the signs of a and b and also whether a is larger or smaller than b in absolute value. She has no mental image of what she is doing but simply applies the eight rules she has learned. She succeeds about two-thirds of the time—enough to pass any short-answer standardized test.

However, Marion can do well with certain other problems that can be regarded as purely algebraic in nature. An example follows:

Teacher: [Writing on the chalkboard, he says the following.] Three apples and 2 bananas cost 13 cents. Two apples and one banana cost 8 cents. What is the price of 5 apples and 3 bananas?
Marion: I don't know. How much does an apple cost?
Teacher: I'm not sure. But I don't think you need to know. Think of a package of 3 apples and 2 bananas. [He draws on the board.] And then a package of 2 apples and 1 banana. [He draws on the board.]
Marion: Oh. You just get one of these and one of those.
Teacher: So how much do you spend?
Marion: $13 + 8 = 21$ cents.
Teacher: And suppose you wanted to get 6 apples and 4 bananas?
Marion: Easy. You buy 2 of those [she is pointing to the first package] for 26 cents.
Teacher: And how about 8 apples and 5 bananas?
Marion: [She thinks a bit.] You could buy 4 of those. [She points to the package of two apples and a banana.] No. That would only give you 4 bananas. We need 5. I give up.
Teacher: Well, take some of these [he labels the package of 3 apples and 2 bananas as "P"], and some of those [he labels the package of 2 apples and 1 banana as "Q"].
Marion: OK. I could take 2 P's. Then I need 1 Q to make 8 apples.
Teacher: But do you have the right number of bananas?
Marion: Well... yes! I get 4 bananas from the 2 P's and 1 more from the Q.

Discussion of Vignette 2

This lesson evolved into a unit on the solution of simultaneous equations. Later on, the students were able to use the letter a to represent the unknown price of an apple and b to represent the price of a banana. They then were able to add and subtract linear expressions in a and b. They also could "scale" them by writing $5(3a + 2b) = 15a + 10b$, for example. Often, they had to return to the drawings of packages of apples and bananas for inspiration.

Perhaps more important, the students in this class were able to perform the quintessentially algebraic operation of substitution. That is, their description of $5(3a + 2b) + 3(2a + b)$ as "five of these packages and three of those packages" evolved into the expression 5P +3Q, where P stood for $3a + 2b$ and Q for $2a + b$.

Marion and her classmates were learning useful algebraic operations without referring to arithmetic. That is, they would have more trouble evaluating $3a + 2b$ when $a = 1/5$ and $b = 2/3$ than they had evaluating 5P + 3Q as described above.

Conclusion about the Vignettes

The two vignettes above are intended to show examples of how algebraic thinking can be distinguished from arithmetic, or generalized arithmetic.

HOW IS THE STUDY OF ALGEBRA RELATED TO THE STUDY OF FUNCTIONS?

The confounding of a function with its algebraic representation has deep historical roots. It is common in the history of mathematics to find that the idea of a function originated as a name for an algebraic expression.[11]

From a more modern viewpoint, it is clear that the study of functions is not the same as the study of algebra. Many functions can be represented graphically more easily than algebraically. For others, we cannot write a closed algebraic expression and so have contrived names, such as $\sin x$ or $G(x)$, to remind us of the function of which we are thinking. In still other cases, such as functions of a random variable, the functional values are not related to computations, either arithmetic or algebraic. In each of these cases, however, we use techniques of algebra to handle the functions, once they have been named.

It is equally clear that algebraic expressions can be studied apart from their use in representing functions. Perhaps the simplest example of this is in studying the integers modulo N (for some fixed N). Two polynomials with coefficients in such a set may be distinct as polynomials but may define the same function.

Having separated algebraic thinking from thinking about functions, we should expect that the corresponding pedagogy might differ as well. Indeed, while algebra is a useful tool for the study of functions, recent literature is replete with examples of classroom work supporting the function concept in ways that avoid the use of algebraic expressions.[12]

AN HYPOTHESIS CONCERNING THE TEACHING OF ALGEBRA

To summarize the argument so far:

(1) Algebraic thought begins with the study of the field of rational expressions;
(2) Algebraic thought is not the same as working with variables nor is it the generalization of arithmetic nor is it the study of functions; and
(3) Algebraic techniques find important applications in generalizing arithmetic and in studying functions.

What should we make of the experience of so many teachers (including the author) that a direct study of algebra as a set of formal rules for manipulating rational expressions leaves the student unenlightened? Classroom practice has shown that a frontal attack at the ideas of algebra (as defined by Gelfand's remark) will not give the student access to the power and utility of algebraic techniques but, rather, will render it, as one colleague has said, "an intensive study of the last three letters of the alphabet."[13]

---

[11]For example, Euler sometimes used the term function this way. See example, Carl B. Boyer, *A History of Mathematics* (New York: John Wiley and Sons, 1968), 484ff.

[12]See, for example, Stephen S. Willoughby, "Activities to Help in Learning about Functions," in *Mathematics Teaching in the Middle School*, Vol. 2, No. 4, February 1997. The author of this article makes a conscious and successful attempt not to use algebraic expressions in addressing the function concept with middle-school students.

The mathematical statement of Gelfand concerning the relationship of arithmetic to algebra perhaps can be used as a guide here.

It has long been recognized that the learning of arithmetic is most effectively approached through its applications. Early learning about numbers must be very concrete: a set of two birds has something in common with a set of two cupcakes, and both are different in a certain way from a set of three cupcakes. Later on, this "something in common" becomes the concept of cardinal number. Similarly, teachers build on concrete applications of the operations of addition and multiplication to develop the concepts they imply, then the algorithms that reflect them.

Our hypothesis is that the learning of algebra proceeds in just the same way. We start students on the road to algebra with its concrete applications, such as the description of number patterns or functions. Later on, the algebraic ideas are used to solve polynomial equations (where the algebraic expressions represent rational, real, or complex numbers) and to manipulate functions. Slowly, through years of experiences, the algebraic expressions take on a life of their own, and the student can use them wherever necessary without reference to their interpretation as numbers, for example. The student is then working within the field of rational expressions.

A COURSE CALLED ALGEBRA

A look at the "algebra" taught in schools today confirms this hypothesis. We have slowly moved from a stiff and formal course in symbolic manipulation towards an approach to algebra through its applications. Indeed, the typical first-year "algebra" course contains very little algebra (in the twentieth-century sense). It has long included such arithmetic topics as signed numbers and numerical radicals (although a case can be made that $\sqrt{3}$ is neither treated nor perceived as a number, in the sense of quantity, in these courses). For at least 40 years, this course has included extensive work with functions, including graphical and other representations, as well as the use of algebraic techniques. In more recent decades, it has included work in combinatorics and even logic, both with and without the application of algebraic concepts to these areas.

This is not a new observation but only a new description of an evolutionary process that we have taken in matching our pedagogy to the achievements of twentieth-century mathematics. It explains and unifies a number of recent observations about the teaching of algebra. For example, Jim Kaput calls for us to "de-coursify Algebra: to weave strands of algebra throughout the grades, so that algebraic ideas grow naturally with the child."[14] This practice has proved effective in the teaching of arithmetic. There is no "course" labeled arithmetic in elementary school. The teaching of arithmetic has been integrated into the development of other mathematical concepts. Perhaps the most emblematic acknowledgment of this change was the renaming of the National Council of Teachers of Mathematics journal from *The Arithmetic Teacher* to *Teaching Children Mathematics*.

The danger in "de-coursifying" algebra is that we may lose what is valuable in the subject. Identifying the study of the field of rational expressions as the essential beginning of algebra will help us keep algebraic thinking in the curriculum, while we also approach it through other areas of mathematical thought.

ALGEBRA FOR ALL?

How do these observations about the nature of algebra impact on the issue of who should learn algebra? Is it important for lawyers to learn about the field of rational expressions? Or nurses? Or kindergarten teachers? Why should the study of rational expressions be so important to the work force of tomorrow?

These are not simple questions. A deeper consideration, which is out of place here, would quickly lead us close to a central problem of the philosophy of mathematics; namely, the question of why the mental constructs of mathematics (if that is what they are) so unreasonably match our experience of the physical world. The connection

---

[13]Beverly Williams, personal communication with the author, April 1997.

[14]Jim Kaput. "Long-Term Algebra Reform: Democratizing Access to Big Ideas," in *The Algebra Initiative Colloquium*, edited by Carole B. Lacampagne, William Blair, and Jim Kaput (Washington: U.S. Department of Education, 1995), 34.

between the learning of mathematics—any mathematics—and physical (i.e., economic) survival in later life is more complex than some recent writings would indicate.

A simplistic answer, sometimes given by educators whose expertise is in fields other than mathematics, is a sort of "socialist realism" doctrine of mathematics: that school mathematics must always be related to concrete physical objects or demonstrably useful in earning a living. This solution is certainly not correct. One can think of numerous specific exercises or problems that may seem contrived (from this point of view) but that serve as efficient stepping-stones to a deeper understanding of mathematics. Algebra (in Gelfand's sense) may be seen as one of these stepping-stones.

In another sense, a clear understanding (on the part of teachers and writers of curriculum) of algebra and what it means can be seen as an important unifying strand for the teaching of other, more obviously "useful" mathematics. Finally, offering algebra to all students is offering an opportunity to all students. Not many of our students will end up thinking like Emmy Noether, but it is important to allow as many students as possible access to this career path, whether or not they eventually follow it.

For all these reasons, twentieth century algebra, like other areas of mathematics, is certainly important for all students—whether or not the students themselves reach the point where they can consciously identify the algebraic ideas they are using.

ALGEBRA AND TECHNOLOGY

We have seen that the focus of algebraic thinking is on the study of the field of rational expressions and operations on them. How have, or how will, the use of new technologies impact the study of algebra? We can do no more than suggest where to look for fruitful applications and see how these opportunities play out in future classrooms.

We can distinguish two categories in the use of new technologies. On the one hand, technology can allow us to teach better (or sooner) concepts that we have been teaching all along. On the other, technology can allow us access to new concepts or ideas that we have not been able to address in more traditional environments.

But in beginning this search for opportunities, we immediately run into a difficulty: What is "better?" Does sooner mean better? Does easier mean better? Which uses of technology are of value, and which are window dressing?

Again, our examination of the nature of algebraic thinking can help out. We already have a certain amount of experience with the use of calculators in teaching arithmetic. Perhaps we can anticipate some of the same experiences in teaching algebra.

For example, a judicious use of the calculator can help with the acquisition of mechanical arithmetic skills. By anticipating the calculator's results, students can deepen their sense of how the operations of arithmetic work. By exploring number patterns with the calculator, students can begin the process of generalizing arithmetic much earlier and much more naturally than they could without calculators. By using the calculator with the ugly numbers of "real" data and comparing their work with a problem involving simple but contrived data, they can strengthen their concept of the arithmetic operations.

Let us look at the corresponding situations for algebra. One algebraic analogue of the calculator is symbolic manipulation software, such as "Derive," "Mathematica," or "Maple" (although symbolic manipulation software is closer to algebraic computation than most calculators are to arithmetic computation because the issue of decimal representation and approximation of answers does not occur).

On the simplest level, students can certainly check algebraic manipulations against the software. But the possibility of doing algebraic computations quickly and accurately immediately presents another opportunity; of exploring algebraic patterns.

The following is an example of a classroom exploration that I have found useful where good use of symbolic manipulation software can be made. This exploration can be used as early as a first-year algebra level or with more advanced students. We will look at how the problem set might be used without, then with, symbolic manipulation software.

Problem 1: Factor $x^2 - 1$
(Students who cannot do this easily are not ready for this exploration.)

Problem 2: Factor $x^3 - 1$
(We assume that students are not yet familiar with the factorization of the difference of two cubes. If students are working the examples by hand, it might be best to ask them to show that $x - 1$ is a factor of $x^3 - 1$.)

Problem 3: Factor $x^4 - 1$

Problem 4: Factor $x^5 - 1$
(By this time, students will generally be able to guess that $x - 1$ is a factor. The form of the other factor is not difficult to guess in this particular case. Or, students can use a division algorithm for polynomials to obtain it.)

Problem 5: One of two factors of $x^{15} - 1$ is $x - 1$. What is the other factor?

(Students frequently are able to guess the form of the second factor, and checking the computation by hand in this case is not really tedious.)

From here, the exploration can take a variety of paths. Two examples are the factorization of $a^n - b^n$ in general and the formula for the sum of a (finite) geometric progression.

How does the same sequence of problems play out with symbolic manipulation software at hand? The first four problems are not very different. However, the software will probably give the complete factorization of $x^n - 1$ and not just the two factors that indicate the particular pattern shown above. If students guess at the form of the second factor, they can check it with the software. This guess can be motivated by the form of the second factor when n is prime.

But much more can be done with the software. For example, students might consider the following factorizations, which would probably not be available to them working manually:

$$x^5 - 1 = (x - 1)(x^4 + x^3 + x^2 + x + 1)$$

$$x^{10} - 1 = (x - 1)(x - 1)(x^4 + x^3 + x^2 + x + 1)(x^4 + x^3 + x^2 + x + 1)$$

$$x^{15} - 1 = (x - 1)(x^2 + x + 1)(x^4 + x^3 + x^2 + x + 1)(x^8 - x^7 + x^5 - x^4 + x^3 - x + 1) \text{ (and so on).}$$

A variety of patterns can be observed, the most prominent of which is that certain factors keep coming up. A bit of inquiry will yield the relationship between these recurring factors of the polynomial $x^n - 1$ and the prime factors of the number n. While a complete study of the cyclotomic polynomials is certainly outside the scope of a high-school course, many insights can be gleaned on this level. Some of these can be developed into methods or theorems, especially with the study of complex numbers (for which an entirely different set of software explorations might be developed) or the factor theorem. Work with algebraic patterns is not yet well documented in the literature. Some easy places to look for applications of this idea are in exploring the binomial theorem, teaching factoring patterns, or exploring algebraic symmetry (for example, by using the software to permute the variables). An entirely different set of activities allows for the exploration of algebraic ideas through their applications in studying functions.

CONCLUSIONS

The current trend in education is to unify the teaching of different areas and to create strands throughout the grades of subjects formerly taught in discrete packages. This trend has helped us to reach past mathematical techniques to mathematical concepts. As we do this, it is important that we not lose track of the nature of the concepts we are addressing. An inquiry into the nature of algebraic thinking and its relation to other areas of mathematics, as well as into the ways in which we encourage it by our students, will help us in our efforts to place this important tool at our students' disposal.

Appendix E

A Framework for Constructing a Vision of Algebra: A Discussion Document

Working Draft

This document has been adapted from the "Algebra in the K-12 Curriculum: Dilemmas and Possibilities," submitted in March 1995 by the Algebra Working Group to the National Council of Teachers of Mathematics. This document reflects the comments and suggestions on the original document by the National Council of Teachers of Mathematics Board of Directors and reviewers from the mathematics community. Permission to photocopy materials from this document is granted to individuals and groups who want to use it for discussion purposes.

May, 1997

The National Council of Teachers of Mathematics, Inc.
1906 Association Drive, Reston, VA 22091-1593

Printed in the United States of America

NATIONAL COUNCIL OF TEACHERS OF MATHEMATICS
THE ALGEBRA WORKING GROUP

Gail F. Burrill
University of Wisconsin-Madison
Madison, Wisconsin

Jonathan Choate
Groton School
Groton, Massachusetts

Joan Ferrini-Mundy
University of New Hampshire
Durham, New Hampshire

Steven Monk
University of Washington
Seattle, Washington

Beatrice Moore-Harris
Fort Worth Public Schools
Fort Worth, Texas

Elizabeth Phillips
Michigan State University
East Lansing, Michigan

Merrie L. Schroeder
Price Laboratory School
Cedar Falls, Iowa

Jacqueline Stewart
Okemos Public Schools
Okemos, Michigan

Lee V. Stiff
North Carolina State University
Raleigh, North Carolina

Erna Yackel
Purdue University-Calumet
Hammond, Indiana

Mary M. Lindquist, Board Liaison
National Council of Teachers of Mathematics
Reston, Virginia

CONTENTS

PREFACE

In 1994, the National Council of Teachers of Mathematics Board of Directors created an Algebra Working Group and charged it to

> produce a document that: *Expands the vision of algebra for all that begins with experiences in early elementary school and extends through secondary school; elaborates this vision by including example, practical ideas, and promising practices, and helps school systems raise questions about the process of change.*

The Working Group met in the summer and fall of 1994 and developed a draft document that was circulated broadly within the mathematics, mathematics education, and school community for review and comment. This document entitled *A Framework for Constructing a Vision of Algebra* was presented to the National Council of Teachers of Mathematics Board of Directors in 1995 as a final report of the Working Group. It has been circulated since that time upon request to many groups and individuals interested in questions about school algebra.

In anticipation of the May, 1997 National Council of Teachers of Mathematics (NCTM)/Mathematical Sciences Education Board (MSEB) symposium on the nature and role of algebra in the K-14 curriculum, the document has been revised and updated to serve as a discussion and background document for the symposium.

Gail Burrill
Joan Ferrini-Mundy
Algebra Working Group Members
May, 1997

INTRODUCTION

The release of the National Council of Teachers of Mathematics' *Curriculum and Evaluation Standards for School Mathematics* (NCTM, 1989) marked a new era in K-12 mathematics education. The Curriculum Standards call for a rethinking of the mathematical goals and emphases of school mathematics. In particular, the document outlines ways in which the subject matter of algebra can be organized as a strand occurring throughout the K-12 grade span, rather than confining algebra to the typical two courses in high school. The Patterns and Relationships standard for grades K-4, for example, calls for students to

- recognize, describe, extend, and create a wide variety of patterns;
- represent and describe mathematical relationships; and
- explore the use of variable and open sentences to express relationships (NCTM, 1989, p. 60).

The standards for grades 5-8 include a standard called Patterns and Functions and an Algebra standard. For grades 9-12, there is an Algebra and Functions standard as well as a Mathematical Structure standard. Taken as a whole, these five standards offer one sketch of a K-12 algebra strand.

Curriculum developers, textbook authors, and others have elaborated such conceptualizations of K-12 algebra in their publications and materials. The NCTM through its publications (*Algebra for Everyone;* the Addenda Series for grades K-6, *Making Sense of Data and Patterns;* for grades 5-8: *Dealing with Data and Change* and *Patterns and Functions;* for grades 9-12: *Algebra in a Technological World* and *Data Analysis and Statistics across the Curriculum;* and the February, 1997 special issues of *Teaching Children Mathematics, Mathematics in the Middle Grades,* and the *Mathematics Teacher*) also has provided further elaboration and discussion of how a K-12 focus on algebra and algebraic thinking might be formulated.

This document contributes further to the ongoing examination and work of shaping the school algebra curriculum, largely through a proposed framework for organizing discussion about algebra in the K-12 curriculum. We also offer extended examples of how algebraic reasoning might be developed and encouraged across the grades.

A number of current pressures contribute to the need for *ongoing* examination of the algebra dimension of the school curriculum. Problems in the workplace, in industry, and in everyday life involve algebraic concepts. Fundamental mathematical ideas in the areas of growth and change, patterns and regularity, quantity, size, shape, and data are often expressed with the tools and symbols of algebra.

Increasingly sophisticated technology opens a wide range of possibilities of rethinking the emphases that have been traditional in school algebra, and raises a set of serious questions. Fundamental issues about the type and amount of symbol manipulation and procedural activity that is appropriate for students can now be examined and debated within the context of heretofore unavailable technological tools. New conceptualizations of "symbol sense" (Arcavi, 1994) and "function sense" (Eisenberg, 1992) have emerged within the general discussion of algebra teaching and learning. In addition to considering how the available technologies might be used to help students understand the concepts of algebra and the procedures of algebra, there is now the dual question of how these technologies might themselves necessitate changed emphases and new additions to the content of school algebra.

Research and practice provide compelling evidence that children engage in significant mathematical reasoning at early ages, and that algebraic thinking can be nurtured and encouraged early in the curriculum (Bastable & Schifter, in preparation; Kaput, in preparation). How can the school algebra curriculum be formulated to develop across the grades and to capitalize on these understandings? What might such early introduction of key algebraic concepts and processes mean for the revision of the traditional secondary school algebra curriculum?

Currently, there is a strong trend toward algebra for all in the nation's eighth grades and secondary schools. Yet, various sources of evidence indicate that, for many students, their experiences with algebra in middle and secondary schools are not leading to high levels of understanding or proficiency (Beaton et al., 1996; Reese et al., 1997). Compounding the situation, curriculum and instructional materials currently available provide a wide, and sometimes confusing, array of distinct possibilities of how the algebra curriculum might be organized.

The context and climate around algebra as a K-12 element of the mathematics curriculum is ready for discussion. The framework and examples that follow are intended as a contribution to the process of continuing deep discussion about this important area of mathematics education.

A PROMISING PRACTICE

Many people assume that "algebra" means working with symbols, but in recent years there has been a great deal of discussion among the mathematics education research community on reasoning that promotes understanding of important algebraic concepts at early levels (Bastable & Schifter, 1997; Confrey, 1995; Confrey & Smith, 1995; Harel & Confrey, 1994; Thompson, 1995). The following discussion parallels an episode that occurred mid-year in a heterogeneous sixth grade class.* The students were studying a unit on rational numbers, and the intent of the problem was to develop understanding of and methods for comparing ratios. The problem illustrates how students' reasoning can be used to develop understanding of algebraic ideas.

Description of the Problem: The school is hosting a lunch for the senior citizens in the area, and a class is asked to test four different recipes for mixing punch made from sparkling water and cranberry juice. First, they have to decide which of the four has the most cranberry juice, and then determine how many cups of juice and water are needed to mix 120 cups of punch.

Recipe A: 2 cups cranberry juice
3 cups water

Recipe C: 3 cups cranberry juice
5 cups water

Recipe B: 4 cups cranberry juice
8 cups water

Recipe D: 1 cup cranberry juice
4 cups water

Discussion: The interaction that ensued was very lively. Students interpreted the question of which recipe has the most cranberry juice in a variety of ways. Some looked at the absolute number of cups of juice in each recipe. Others looked at the ratio of juice to water. Still others used the part-whole relationship of the number of cups of juice to the total number of cups in the recipe. After comparing methods and discussing choices, the class split into four groups to decide how to adapt each recipe to make enough punch for 120 cups. The groups reasoned as follows.

*This vignette is based on an episode from a sixth grade class taught by Mary Bouck, who was piloting materials from the Connected Mathematics Project (CMP), a National Science Foundation funded middle school mathematics curriculum project (Fey et al., 1995).

Group With Recipe A: We thought each recipe would make five cups: 2 of juice and 3 of water. So to make 120 cups, it would take 120 divided by 5 or 24 batches of our recipe. To get the number of cups of juice, multiply 2 by 24 to get 48 cups of juice. To get the number of cups of water, multiply 3 by 24 to get 72 cups of water. So 48 cups of juice plus 72 cups of water gives 120 cups of punch, and it is still the same recipe because the ratio of juice to water is still 2 to 3 since 48/72 = 2/3.

Group With Recipe B: We saw there were twice as many cups of water as juice, so we divided 120 cups into three equal parts: 40 + 40 + 40 cups. Since there is twice as much water, two of the parts must be water. So there are 40 + 40 or 80 cups of water and then 40 cups of juice. This gives 120 cups of punch, and the ratio of juice to water is still 1/2.

Group With Recipe C: It took us a long time. We tried to double the recipe, and this was not enough. We added another batch, and this still did not get us 120 cups. We finally figured out a pattern: juice/water = 3/5 for 8 cups, 6/10 for 16 cups, 9/15 for 24 cups, 12/20 for 32 cups.... We kept on with the pattern until we got to 120 cups—it is 45/75. We need 45 cups of juice and 75 cups of water to make 120 cups of punch, and the ratio will still be 3/5. (Later, in class discussion, this group observed they could have gone directly from 3/5 to 45/70 by multiplying the "top and bottom" by 15, because they needed to use the recipe 15 times.)

Group With Recipe D: We tried various numbers. First we tried 20 cups of juice. This meant we needed 4 times as much water or 80 cups of water. But this only gave us 100 cups of punch. So we tried 30 cups of juice, which needed 4 times as much water or 120 cups, but this was too much punch. We then tried 25 cups of juice. This needed 100 cups of water, but this made 125 cups of punch, which is close but too much. So we tried 24 cups of juice, which needed 96 cups of water. This worked! We had 120 cups of punch, and the ratio of juice to water is 1 to 4. (One of the students noticed that the 125 cups contained one too many batches of the recipe, so she subtracted one batch, that is she took away one cup of juice and 4 cups of water. She came up with the same amount as the rest of the group—24 cups of juice and 96 cups of water.)

During the discussion directed by the teacher, the class thought about the similarities and differences among the strategies used by each group. Some concluded that the number of each recipe needed for 120 cups of punch could be found by using the multiple necessary to obtain an equivalent ratio

$$\frac{\text{cups of juice}}{\text{total cups of punch}}$$

with a denominator of 120.

Where Is the Algebraic Reasoning?

While the students did not use symbols, they did develop their own representational schemes to organize their thinking and build a sense of variable. A closer examination of the reasoning process reveals that students were also developing understandings of proportions, rates of change, and linear relationships.

Recipe A: This group seemed to have the following organization in mind:

Cups of Cranberry Juice	Cups of Water	Total Cups Punch in One Recipe
2	3	5
?	?	120

By reasoning about quantities, they took any number of recipes, R, and found the total number of cups of punch produced, T_P, using $T_P = 5R$ (a linear equation). For every increase of one recipe, there will be 5 cups more punch (a constant rate of change). They also used this idea to reason about the amount of the individual ingredients; the total number of cups of cranberry juice, T_{CJ}, is equal to the number of cups of cranberry juice in a recipe times the number of recipes, R. Thus, $T_{CJ} = 2R$. Similarly, $T_{water} = 3R$, where T_{water} is the total number of cups of water.

Recipe B: This group also appeared to be reasoning about quantities. They thought of the whole as having three equal parts. Their reasoning focused on the relationship between the cranberry juice (*CJ*) and the water (*W*).

So 120 = two parts *W* + one part *CJ*.

Recipe C: This group created a pattern using pairs of values (or ratios) of cups of cranberry juice, T_{CJ}, and cups of water, T_{water}: (3, 5), (6, 10), (9, 15), … , (T_{CJ}, T_{water}). They inspected the pairs to determine when $T_{CJ} + T_{water} = 120$. (The students solved the equation by reasoning about how the two quantities were related, knowing that $T_{CJ}/T_{water} = 3/5$.)

Recipe D: This group used trial and error to reason about their problem under the constraints:

$$T_{CJ} + T_{water} = 120 \text{ and } T_{water} = 4T_{CJ}$$

T_{CJ} (Total cups of *CJ*)	T_{water} (Total cups of water)	T_P (Total cups of punch)
20	80	100 (too little)
30	120	150 (too much)
25	100	125 (too much)
24	96	120 (right amount)

The students in this class were engaged with a problem that was posed in a context that made sense to them. Allowed to explore the problem on their own, students found several different ways to reason about the problem and the underlying ideas of rate of change, linear relations, and proportional reasoning. In situations such as this it is critical, however, that the teacher recognize and understand the concepts being developed, not, at this level, to teach the abstract manipulation, but to help make appropriate decisions about the conversation and follow-up activities, to focus and direct the discussion in useful ways when the opportunity arises, to use the words that position the algebraic concept (each extra recipe produces 5 cups more punch), and to leave room in students' minds for extending their understandings.

- ***How can similar kinds of experiences that foster algebraic reasoning be created at all grade levels?***
- ***How can such experiences contribute to an understanding of algebra in grades K through 12?***
- ***What knowledge of mathematics and pedagogy does a teacher need to implement rich experiences in algebraic reasoning?***

CRITICAL ISSUES

As different groups address the implications for change, there are debates about the nature, history, future goals, appropriate curriculum, assessment, and pedagogy of school algebra, as well as its relationship to the mathematical discipline called algebra. As these groups focus on research, design curriculum, prepare teachers, teach students, or educate the public, however, several critical issues arise:

- How can algebraic experiences be designed to promote success for all students?
- What is algebraic reasoning? How is it related to other mathematical reasoning?
- What are the implications of the differing views of school algebra?
- Should algebra be a course?
- How can coherence be brought to a K-12 algebraic experience?
- What is the role of symbols in a K-12 algebraic experience? What is the relation between the development of conceptual understanding and acquisition of algebraic skills? How much symbol manipulation is needed to develop symbol sense?
- What knowledge about algebra is needed to use technology effectively? How does technology enhance algebraic reasoning? What other tools can help?
- How will algebraic reasoning and algebraic understanding be assessed?
- What teacher preparation is needed to implement a coherent algebraic experience in a K-12 curriculum?
- What are the implications of state-mandated algebra courses?
- What is the relationship between school algebra and research areas of mathematics?

While these issues are not meant to be exhaustive, they are representative of those that commonly arise when educators consider the nature of algebra and what it should encompass. Some of these issues have been discussed in earlier works of the Council, in particular, *Research Issues in the Learning and Teaching of Algebra* (1989) and the 1988 yearbook, *The Ideas of Algebra K-12*. Many of the issues are common to mathematics in general as well as to algebra. Changes in the algebra curriculum cannot take place in isolation from the rest of mathematics, and in fact, rethinking algebra is only one phase of rethinking the entire K-12 mathematics sequence. If the suggestion that algebra is a K-12 enterprise is realized, the results will force a reconceptualization of the scope and sequence of the entire mathematics curriculum. A rich K-8 algebra experience will be of little value if students are placed in an algebra course that ignores the knowledge and understanding they bring from such a background. These changes can, in fact, be observed in the large-scale curriculum projects funded by the National Science Foundation. Some of the issues have been the object of prior research and investigation: pedagogy (Rachlin, 1982); interpreting graphs (Russell et al., 1995); quantitative reasoning (Thompson, 1995; Tierney & Monk, in press); formal symbolism (van Reeuwijk, in press). Other issues, particularly the relationship between the development of conceptual understanding and acquisition of algebraic skills, are part of a much needed research program. It is not the intent of this paper to address each of the issues, but rather to provide a framework for thinking about algebra that can be used as educators wrestle with their own responses to the issues. The paper describes differing views of algebra, argues that algebra should be more than a course, and gives some examples that illustrate how algebra might look in a K-12 setting.

These issues must be resolved by various groups as they participate in discussions and construct their own vision of algebra for all students in grades K-12.

A FRAMEWORK—
BUILDING A DYNAMIC VIEW OF ALGEBRA

Algebra has its roots in a variety of human activities, from commerce to the study of numbers, to the elaboration of other branches of mathematics. The centrality and importance of algebra in mathematics reflects its power to express compactly and efficiently mathematical relationships within these activities and to reveal structures and patterns common to them. The usefulness and vitality of algebra are generally seen as arising from its concepts and its patterns of reasoning as they have been adapted in the attempt to solve problems in ongoing human activities. A framework for thinking about algebra should capture the notion that "algebra is a symbol system of unparalleled power for communicating quantitative information and relationships; it is a training ground in careful rule-governed reasoning; its development is a significant thread in the history of mathematics; and its theoretical structure is based on concepts and principles that generalize to provide organizing schema in nearly every other branch of mathematics" (Fey, 1989, p. 207).

Such a framework might be thought about in two parts: contextual settings and organizing themes. The Punch Problem provides a context for algebraic thinking, and it is important to have some organizing theme to develop this thinking across the grades. The development of algebra and algebraic reasoning is embedded in contextual settings; these contextual settings are connected by organizing themes that bring coherence to the curriculum.

Embedding Algebraic Reasoning in Contextual Settings

Historically, algebra was primarily the domain of an elite where only the privileged took part. It essentially consisted of solving equations using problems that were artificial and until the 18th century that were based primarily on other topics in mathematics, such as geometry (Katz, 1995). A course called *algebra* gradually worked its way into the high school curriculum during the 19th century. And, until recently, except for a brief time during the era of "new math," algebra was essentially a course or two composed of learning skills needed to solve equations, manipulating symbolic expressions, and solving "word problems." This course was often preceded by "pre algebra," a course that focused on honing students' arithmetic skills in some simple algebraic situations. Views about what algebra should encompass that went beyond this perspective were expressed by some prior to the 1990's, and the advent of technology and symbol manipulation systems forced many others to reexamine the nature of what was called algebra. Common to these views is the belief that the core meaning of algebra extends beyond the routines and procedures of manipulating symbols.

Among the important goals of the NCTM Standards is that mathematics be viewed as problem solving, reasoning, communication, and making connections. Since these thinking and reasoning processes must be *about something*, students' understanding of concepts should be developed through activities embedded in specific **contextual settings.** Contextual settings such as growth and change, size and shape, data and uncertainty, number, and patterns are productive sources to focus mathematical thinking. In particular, this collection of contextual settings, similar to those used in *On the Shoulders of Giants* (NRC, 1990), provides an opportunity for students to learn and to use algebra as well as make connections between algebra and other disciplines. For example, the study of population growth is a context within the setting of growth and change; finding the number of buses used to transport a large group to a fair is a context within the setting of number; and deciding how to establish a formula to rank athletes in a given sport is a context within the setting of data. In such settings, the context can give meaning to the representation and allow students to recreate the meaning when necessary for understanding.

These contextual settings also help students organize and interact with the world around them: ideas like growth and change can be realized in contexts as varied as a growing bacteria colony, a moving car, or expanding a recipe for punch. The commonality among these situations is how change in one quantity is related to change in another. It is critical, however, to respect the nature and structure of the context and the relationship to the underlying mathematics.

Contextual settings can be based on applications, mathematics itself, or mathematics from a historical perspective. Describing the relationship between the zeros of a polynomial, its factors, and its graph is an exploration within mathematics. Generating an algebraic analog of the geometric reasoning and conclusions made by early mathematicians such as Fibonnaci can be fascinating for many students. Trisecting an angle or constructing a circle with a given area using a straight edge and compass are two historical geometry problems that led to a need to find solutions to polynomial equations.

Students build concepts and develop ways to think in pursuit of activities that engage them in different contextual settings; such settings help students make sense of the algebra they are studying.

Bringing Coherence to the Algebra Curriculum—Organizing Themes

Contextual settings help bring meaning to important mathematical ideas. There is an intrinsic difficulty, however, in any attempt to build a mathematics curriculum out of students' experiences with a range of concrete contexts, in that such a curriculum may be perceived by students as essentially disconnected and not coherent. If a student is deeply immersed in one stimulating contextual problem and then another, this student may not see ideas common to both problems. Even though teachers or curriculum designers might see particular thinking patterns or concepts common to a range of problems, there is a danger that students might not see any of these commonalities because of the very richness of the problems themselves. In the promising practice problem involving recipes for punch, students were thinking about relationships between quantities. Yet thinking about unit rate, proportionality, or linear functions can connect this problem to other problems that at first appear to be of a very different nature. It is the ability to generalize, extend, apply, and connect ideas throughout many different situations and across several grades that enables students to make sense of mathematics and to think of it as a set of unifying ideas rather than a set of disjointed, unrelated problems.

One of the ways in which the various efforts at reforming algebra differ most strongly is in the **organizing themes** or conceptual organizers that have been selected to give coherence to the subject by those developing curricula, implementing curriculum, or conducting research. As school algebra evolved over time, the themes used to organize the subject have changed. The "new math" of the 60's was based on structure. This was followed by a "back to the basics" movement, based heavily on thinking of algebra as a language. In the 90's, with the emergence of the graphing calculator, the concepts of function and modeling have become central points for conceptualizing algebra. To be effective in helping students make connections and understand the big ideas, teachers and curriculum writers must allow a variety of themes to emerge.

Some of the themes that have been often discussed and realized in written curricula are described below.

- ***Organizing Theme: Functions and Relations***

  The central mathematical concept of a relationship underlies the equations, tables, and graphs so common to algebra. Representing a relationship using the concept of variable provides organized ways to think about an enormous variety of mathematical settings. The study of functions as the focus of algebra has been advocated by Fey and Good (1985) and Yerushalmy and Schwartz (1991). Functions can be expressed in virtually all of the representational systems found in algebra and can serve as the basis for exploring interesting problems. In studying population growth, students might represent the function with a table, graph, or symbols. In a situation involving a geometric transformation, students might represent the transformation with matrices. Recursive definitions of functions, because of their importance in computer methods, play a role in problems in science and economics.

Change and variation are important ideas within the concept of a function and can be realized in contexts throughout the curriculum. Examples such as what happens if a pizza must be shared among larger and larger groups of picnickers are appropriate for elementary students. In high school or college, examples might include the study of the derivative of a function and its applications. The relationship between the amount of cranberry juice and sparkling water used to make punch is a direct proportionality, a special case of the larger family of linear functions.

There are other situations where a relation describes the situation. Consider, for example, the Pythagorean Theorem that relates the lengths of the sides of a right triangle or Euler's Formula that relates the number of vertices, the number of edges, and the number of faces of a geometric figure. In the elementary grades, students could look for number relationships in their class, such as "the number of noses and ears is equal to three times the number of students in the class." The ideas that naturally arise in the contexts noted above can stimulate valuable mathematical activities for students in virtually all grade levels. At the same time, these activities can all be seen as fitting within the development of a function-and-relationship theme.

- ***Organizing Theme: Modeling***

Modeling as a theme for thinking about algebra is evident in the work of Freudenthal (1983) as well as others. Many complex phenomena can be modeled by relatively simple algebraic relationships. Conversely, the phenomena can be seen as embodiments of the relationships, so that engaging in the complex processes of modeling the phenomena will lead to grappling with the relationships. In the early grades, students can investigate the relationship between the number of shoes and the number of people by using counters to represent the number of shoes. Reasoning with these quantities helps children find patterns and make predictions about the number of shoes and the number of people. Middle grade students can explore the relationship between the stretch in a bungee cord and the weight of the jumper using a rubber band and fishing weight. By varying the factors in the experiment and recording the results, students find that an underlying linear relationship is present. Older students can generate the symbolic form for the relationship, look at the limits of the situation compared with the model, and create a spreadsheet to compare the predictions with the actual results.

In other situations, such as the amount of a prescription drug present in the body, recursion relationships can be used to model the amount of drug in the body, taking into consideration the dosage, metabolism rates, and other factors. In yet other situations, matrices or systems of equations can be used to model ecological networks or economic decisions. The power of a model lies in its accessibility and utility. A simple function, such as $f(x) = x(a - bx)$, might represent two entirely different situations: the area of rectangles with a fixed perimeter where $P = 200 = 2L + 2W$ and $A = L(100 - L)$ or the revenue collected when selling a product for which the demand depends on the price. If the demand is given by $D = 200 - 3p$, where p is the price, then the revenue is given by $R = p(200 - 3p)$. Knowledge gained in one context can inform a totally different application of the same model. Viewing algebraic relations in terms of the phenomena they model is an effective way of giving life to the situations and bringing to the study of algebra the richness of experience all students carry with them.

- ***Organizing Theme: Structure***

Through the efficient and compressed symbol systems of algebra, deep yet simple structures and patterns can be represented. It does not matter in which order a column of figures is added; it does matter in which order two numbers are subtracted. This is often a source of puzzlement to students in the early grades. By examining these phenomena across many number systems and many operations, the commutative property emerges as a way to capture the essential nature of the systems. One of the most puzzling features of matrix multiplication is that it is *not* a commutative operation, yet a characterization of the formal properties of matrix multiplication can be used to manage a wide variety of problems.

For some, structure is the core of school algebra (Kiernan, 1989; Cuoco, in press), building on generalized arithmetic, the formulation and manipulation of general statements about numbers. This approach requires students to move away from context-dependent problems and to develop the ability to generalize and follow procedures in a systematic way. Organizing algebra around structure implies thinking about how systems operate; for instance, what enabling characteristics of a system allow fractions to be combined or equations to be solved? The power of mathematics lies in its ability to abstract and generalize the common features from a system and then to apply and extend these ideas to other systems. The search for primes by looking at the decomposition of whole numbers reveals information about the structure of the real numbers. Understanding the structure of integers allows us to seek and recognize similar structures in other systems, such as the ring of polynomials: the set of integers and the set of polynomials have a similar algebraic structure. The study of structure is a means of focusing on common aspects of many mathematical situations; a consideration of structure allows us to look beneath the more superficial aspects of the particular notational system and beyond the concepts arising from particular contexts.

- ***Organizing Theme: Language and Representation***

Algebra can be seen as a language—with "dialects" of literal symbols, graphs, tables, words, diagrams, and other visual displays. Thinking about algebra as the "language of arithmetic" has been an ongoing part of the discussion (Usiskin, 1988). Euler's 1757 *Introduction to Algebra* summarized all of the previous work on solving equations, but used no diagrams; his analysis consisted of the manipulation of numerical quantities. Although that view of algebra still permeated the majority of the content in algebra texts for most of the 1900's, the graphic capability of computers and calculators has opened the door to new and easily accessible geometric and tabular representations. The relation between $y = x^2 - 4$ and $y = (x - 2)(x + 2)$ and the corresponding graphical interpretation enables students to picture concepts in new and enriching ways. A discussion of the information gained and lost as one moves between representations reinforces understanding of the concepts, while changing parameters and observing the impact add a new dimension to student understanding.

Different representations provide students with the ability to express and reconfigure quantities in a wide spectrum of situations. A table of the number of students present in a third grade classroom over a month is a particular way of describing and analyzing the characteristics of that classroom. A graph showing the number of grains of rice given each day in the "King's Chessboard" story enables students to describe and analyze the characteristics of that interesting situation.† The same idea lies behind the use of spreadsheets to analyze a number of numerical phenomena in more complex situations. Developing ease in setting up and manipulating graphical and tabular representations of data, regardless of the particular situation being represented, can be a source of considerable mathematical power for students. The same is true of the symbol system of literal variables, expressions, and equations. To represent a situation in familiar terms is often the gateway to understanding the situation. Thus, algebra is a way of thinking about and representing many situations. It has a language and a syntax, along with tools and procedures, that promote this thinking and modeling. By giving students opportunities and time to experience this way of thinking, a more powerful approach to algebra can be built throughout the K-12 curriculum.

- ***What are the essential understandings of each theme?***
- ***How can these themes bring coherence to a K–12 curriculum?***
- ***Are some themes more appropriate at different grade levels?***
- ***Is there a hierarchy to the themes?***

Mathematics is powerful, and algebra is one of the important mathematical tools. No single organizing theme, however, captures the reality of what it means to "know and be able to do algebra." The theme of language and representation, for example, is familiar to many teachers, because it reflects their own school experiences. Its central feature is the use of symbols to translate and represent situations. Students can become fluent in the use of symbols in limited contexts, without realizing that they can be used to model diverse situations, such as maximizing the

---

†In this story, the king rewards his faithful servant by fulfilling his request, which is the amount of rice that can be put on a chessboard in the following way: 1 grain on the first square, 2 grains on the second square, 4 grains on the third square, 8 grains on the fourth square, etc.

number of ships passing through the Suez Canal, determining the amount of medicine in one's bloodstream, or predicting an Olympic record time. In addition, the view of algebra as language and representation has been shifted and broadened by technology. It is now necessary not only to think in the language of algebraic representation but also to extend that language to communicate effectively with computers and calculators. In addition, students who have modeled specific problems, translating and representing important aspects of the problem, can be challenged to raise their understanding to a new level by considering specific strategies, justifying why the strategies work and under what conditions. Learning to ask questions about calculations as they occur across various problems or in isolation from context is an important aspect of algebra—one that leads to a search for common structures. Thus, aspects of language, modeling, and structure complement each other.

Similar reflections on each of the other organizing themes establish that no single theme is sufficient as a base for algebra in the future. These themes are not necessarily parallel or disjoint. However, taken as a collection, the organizing themes can support a dynamic evolution of school algebra. Such a collection provides teachers with multiple entry points into the ways children can think about algebra. In the Punch Problem, the point of entry could be ratio or rates (structure of rational numbers); reasoning about the same situation at a later grade level can lead to students' study of functions or representations. Using a variety of organizing themes will allow more insightful assessment of students' understandings and provide alternatives to help students use algebra. The particular themes highlighted above were chosen because they reflect diverse perspectives, yet have utility and promise for being relevant in many contextual settings. Other themes might be used, but no one organizing theme can generate an algebra curriculum that will adequately serve all students for a multitude of post-high-school options.

Over time, multiple organizing themes should become part of a student's understanding of algebra.

Summary

The "Framework" proposes a way to develop algebraic reasoning by exploring a variety of contextual settings that are connected by organizing themes. By serving as organizers, themes help students recognize important ideas and make connections. Contextual settings are the ground on which these themes play out. They provide the substance from which and about which to reason. The settings and themes suggested above are only a sample and are not intended to be an exhaustive list. Such a framework for curriculum provides opportunities to cross traditional and arbitrary boundaries resulting in a stronger and more useful algebra experience.

The school algebra curriculum must be reconsidered from the ground up, rather than just tinkering with the present curriculum. The challenge is to build a connected and coherent algebra strand by introducing important algebraic ideas at the appropriate grade level and point in the curriculum sequence and to build on this foundation throughout the rest of the K-12 curriculum.

Algebra becomes manifest through exploration of the compelling ideas within the contextual settings, while the themes bring a logic and organization to the discipline.

EXAMPLES FROM CONTEXTUAL SETTINGS

Bringing Meaning to the Framework

The Framework, around which to organize thinking and discussion related to the teaching and learning of algebra, can best be understood by studying examples that bring its meaning into focus. Time spent moving between the general framework and particular examples can result in fruitful deliberations about the intersection of the organizing themes and contextual settings in terms of curriculum, students and their learnings, and teachers and their practice. The discussions are intended to promote continued thinking about the nature of algebra and its role in the K-12 mathematics curriculum.

Each of the following examples was selected to illustrate development of algebraic concepts and reasoning within a particular contextual setting. The problems in the examples show that algebraic reasoning can be developed from the exploration of compelling ideas about our world as well as in mathematics itself. The examples are not in any way meant as a complete catalogue of algebraic topics or as models to be transported directly into the classroom. Rather, they represent important contextual settings for understanding the world around us, generating necessary discussion, and leading eventually to generalization of the underlying concepts. In these examples, ideas of measurement (within the setting of shape and size), exponential growth (within the setting of growth and change), and the distributive property (within the setting of number) are explored. In some examples, one theme may dominate, while in others, several themes may play parallel roles. Some of the problem situations emphasize algebra as a language for the expression and communication of ideas, while in others, algebra as structure helps to further develop understanding of the problem. In still other problems, functions, relations, and modeling are equally prominent. These themes overlap and enhance each other.

A broad view of algebra embedded across grades is needed to make decisions about curriculum and to guide the development of students' reasoning. Thus, the examples not only illustrate how the themes and contextual settings interact within a situation, they also reflect the kind of thinking that may occur in a particular grade level or across grade levels. While each problem is posed initially at one level, it connects to earlier and later levels. When abstract symbols are used, they are not intended for all levels—in many cases they are meant as a possible generalization at a later stage. The reader is expected to play an active role in deciding what age and experience level is appropriate.

To keep the discussion moving, some activities are presented briefly. While reading the examples, it is important to realize that in the classroom:

Students need time to

- explore problems;
- grapple with some uncertainty about ways to solve the problem;

- develop representation systems, even nonhelpful ones, as an important first step in appreciating the power of representation;
- discuss and communicate their reasoning as a means of helping their thinking.

Teachers need to

- assess prior knowledge and experience (or if the activity was part of a curriculum, then some prior knowledge and experience could be assumed);
- facilitate the classroom exploration and discourse;
- provide a balanced and appropriate pedagogy;
- build on experiences and turn difficulties into new perspectives and misconceptions into insights.

The paragraphs with the heading, ***Where Is the Algebra?***, discuss how the example fosters algebra as a system of concepts and a way of thinking. Such paragraphs show how the example might evoke algebraic reasoning and describe how this relates to aspects of algebra familiar to readers. Questions in boxes help readers reflect on concepts, connections, understandings, reasonings, pedagogies, and the choice of context illustrated by each problem. Some questions will also raise issues concerning balance, coherence, use of themes or contexts, or contrasting views. These questions can serve as a basis for discussions about the teaching and learning of algebra. The reader is further challenged to participate in the discussion by reflecting on the following more general questions while reading the examples.

- ***What organizing themes are prominent?***
- ***What other contexts or settings could be used to develop the same ideas?***
- ***What other algebraic concepts and reasoning could be developed using this example?***
- ***Can a similar idea be used at another level of understanding?***
- ***What issues arise? How might they be addressed?***

Example 1: From the Contextual Setting of Growth and Change

Phenomena of growth and change are pervasive in our daily lives. Students bring a wealth of experience, intuition, and curiosity to such everyday events as their own physical growth, the movement of objects in space, patterns of successive change such as repeated doubling, and the growth of a bank account balance with compound interest. Understanding the relationship between two quantities, and in particular how one changes with respect to the other, is critical in areas such as economics, biology, and politics. For example, how fast is the population growing? What is the pattern of growth of the national debt in relation to the size of the population? How does the rate of radioactive decay of certain materials affect our environment? These are important contexts for the development of mathematical ideas. Furthermore, the mathematics developed within these contextual settings can help people understand and make informed decisions about these events.

Situations of growth and change can be represented in a variety of ways: through tables, formulas, graphs, and various types of visual display. Technology has advanced the use of multiple representations, allowing students to investigate the relations among the various representations and to develop a broader understanding of growth and change in its different forms. Each representation can be regarded as a kind of symbol system that has its own particular form of mathematical power for expressing ideas. Eventually, it is necessary for students to be able to use all of these representations, from an informal and personally meaningful representation to the most economical, abstract, and symbolically powerful.

The two problems in this section illustrate how a mathematical idea, in this case exponential growth, can be developed from elementary through high school, thus demonstrating one way to achieve coherence in a K-12 algebra curriculum. The problems, money, and folding paper, also demonstrate the diversity of situations in which a single mathematical idea can be realized. In addition, the themes, *functions and relations*, *modeling,* and *language and representation* are intrinsically involved in working through the problems.

Understanding Growth Using Multiple Representations

Informed decisions about phenomena, such as bacteria growth or the national debt, require an understanding of the essential elements of change. An important part of this understanding occurs when students recognize and represent the fundamental patterns of change. The following problem can begin in early elementary grades and be extended throughout the grades by adjusting the context or the nature of the questions that are asked. The original problem might be posed when students are studying the even/odd property of numbers in the context of folding a piece of paper to form regions. Discovering that the number of regions depends on the number of folds brings the theme of *functions* to the forefront. As students reason about the patterns, they look for appropriate *language and representation* to capture their reasoning.

The Paper Folding Problem: Suppose you want to fold a sheet of paper in half repeatedly to create a region for the name of each person in the class. Only one name may be written in each region—there may be some blank regions. How many folds do you need to make?

Questions to ask: What did the fold line (crease) do to the paper? How many folds did you make? How many regions resulted? How many regions did you have when you had no folds? One fold? Two folds? Three folds? Draw a picture of the paper after each fold. As you go from one fold to the next, how is the number of regions increasing? Does the number of regions increase by the *same* amount each time?

Looking into the classroom: During a preliminary discussion of the problem, one student notices that there is only one region before the folding starts. Other students comment on how the folding is to be done.

The students make a table of the number of folds, the number of regions for each piece of paper, and a sketch of the paper after each fold.

Number of Folds	Number of Regions	Model
0	1	
1	2	
2	4	
3	8	

Looking at their recorded data, students describe the patterns in different ways. One student notices that, after the first, all the numbers are even, but some even numbers, such as 6 and 10, do not occur. Others claim that the number of regions is growing fast. Some notice that the number of regions created doubles each time you add a fold—the number of regions is twice the number of the previous regions. If you fold the paper in half, it cuts every region in half so you have twice as many. Some say that you add 1, then add 2, then add 4, etc.

Where Is the Algebra?

From an organized display of the folded pieces of paper, students can reason about the growth patterns in different ways. Some think in terms of doubling, while others think in terms of adding.

This can lead to a discussion of how these operations are related. Convincing others, for instance, that doubling a number is the same as adding the number to itself can be an important experience in reasoning about numbers. The pattern of repeated multiplication by a given number is a central one in algebra and underlies exponential growth. Finding the number of new regions by multiplying the number of the previous regions by the constant 2 is an example of a recursive process. Students can use a calculator to extend the pattern recursively, and to investigate the increase in the number of regions, from one fold to the next. At the upper elementary level, students can begin to use the terminology of power and exponent in a natural way, moving towards the theme of *language and representation.* Most students will recognize that there are a limited number of folds possible with any size paper. This leads to a conversation about the relation between the pattern and the physical situation. Students might be encouraged to find other situations that have the same pattern but that extend beyond the possible outcomes of folding a piece of paper.

As the discussion continues students are able to focus on the salient features of exponential growth, where the growth pattern is generated by the product of a constant and the previous result. Initially, it is sufficient to express the relationship found using words, models, pictures and either addition or multiplication sentences to represent exponential growth.

While middle school students are studying a unit on growth and change or one on exponents, they might revisit the problem and attempt to generalize by extending the pattern beyond the possible number of actual folds.

- ***How does changing the question from counting the number of regions to analyzing how the number of regions is growing promote an understanding of exponential growth?***
- ***What are the variables in this situation? What are some of the underlying relationships that students might observe and reason about?***
- ***What do student responses indicate about their early understanding of the structure of the number system?***

Paper Folding Problem Extended: Suppose you have a piece of paper that can be folded indefinitely. If you could fold the paper 10 times, how many regions would be formed? Will you have enough regions to write down names for everyone in the school?

Questions to ask: Suppose you could fold the paper 20 times, how many regions are formed? What about 100 times? How many folds will it take to produce 64 regions? 2,048 regions? How about 100 regions? 200 regions?

Looking into the classroom: Since it is impossible to literally continue the folding process beyond a certain number of folds, students look for another way to reason about the growth patterns. Some students extend the table, recursively. Some begin to reason as follows: for 0 folds there is 1 region, for 1 fold there are 2 regions, for 2 folds there are 4 regions or 2×2 regions, for 3 folds there are 8 regions or $2 \times 2 \times 2$. So for 10 folds you multiply 2 by itself 10 times. In response to the second question, some students suggest continuing the table, while others describe the process of multiplying by 2's—20 times. The number of times 2 is used as a factor is equal to the number of folds. For 100 folds, $2 \times 2 \times \ldots \times 2$—100 twos. Some students use tree-like diagrams to show the multiplicative structure.

To find the number of folds needed to produce 64 regions, some students continue to fold the paper; some students extend the table. Others begin to reverse their reasoning patterns for the number of regions: "I got 64 regions by multiplying 2's together, and the number of 2's is the same as the number of folds. So I must divide by 2's until I get to one. It will take 6 folds. For 2,048 regions, it will take 11 folds." To find the number of folds necessary to find 100 regions, some students suggest using a table and observe that 100 is not an entry. It would have to be between 6 and 7 folds, which is impossible. Other students use a graph to find the number of folds. Others reason with the powers of 2: $2^6 = 64$ and $2^7 = 128$. So it is impossible to find the number of folds using the concrete model for "folding in half." Thus, it is impossible to produce exactly 100 regions by repeatedly folding a piece of paper in half.

How does extending the activity beyond the limits of physical folding demonstrate an important aspect of modeling?

Where Is the Algebra?

Some students are ready to discuss patterns in the number of regions without making the actual folds. Others may feel that the certainty of counting is worth more than the labor of repeated folding. At some point, however, most agree that the exploration should be continued without concrete actions. This can be an important first step toward algebraic reasoning about patterns. Noticing, describing, and recording a pattern can lead to two other actions—understanding the mathematics in the pattern enough to make generalizations and using a common mathematical language to describe the situation. Such reasoning also provides understanding and appreciation for a mathematical model.

At this stage, students may use various ways to describe their reasoning. Some use tables.

Number of Folds	Number of Regions	Model	A Shorthand Version
0	1	1	2^0
1	2	1 x 2	$2^1 = 2$
2	4	1 x 2 x 2	$2^2 = 4$
3	8	1 x 2 x 2 x 2	$2^3 = 8$

Some use trees; the opportunity to generalize with words can be a basis for generalizing to a symbolic language at a later stage:

$$R = \overbrace{2 \times 2 \times 2 \times 2 \times \ldots \times 2}^{\text{20 times}}$$

or $R = 2^L$, where R = number of regions, and L = number of lines.

These questions also develop understandings of what it means to solve equations by observing the connections between the various representations. To find the number of regions for 20 folds, students can compute 2^{20} or use the table. To find the number of folds it would take to find 64 regions, students are solving for L in the equation, $2^L = 64$. Students can reverse their reasoning process or use the table. Some students will use a graph to find the number of folds it would take to produce 64 regions or 2,048 regions.

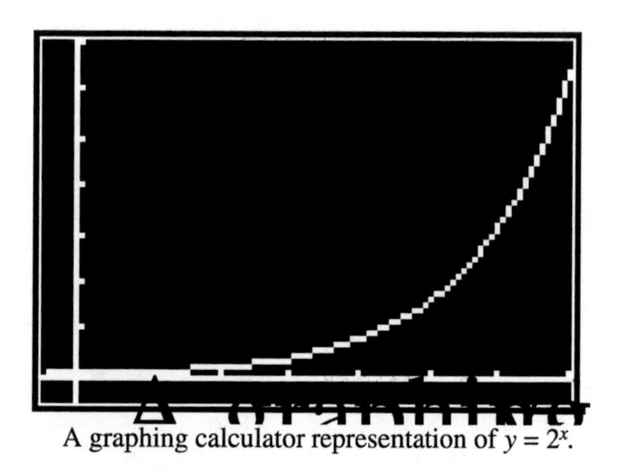

A graphing calculator representation of $y = 2^x$.

If students use a graphing calculator to produce a graph of $y = 2^x$, they may observe that the calculator plots points for non-integer values of x. Students may be asked to consider the meaning of such points and realize that it is possible to fold paper only a whole number of times. Such exploration can encourage deeper thinking about concepts of range and domain and about the limits inherent in any algebraic model of a given situation.

Students should make connections among the entries in the table, points on a graph, and the values of a function. Students can consider how the rule $y = 2^x$ models the situation and the limitations the physical situation imposes. They can think about a way to give meaning to the function for values other than positive integers. They can reason from the input to the output, or in reverse, determine the input that leads to a given output. Asking for the number of folds that would produce 100 or 200 regions also highlights an important characteristic of exponential functions—that the *changes* between outputs grow larger and larger. In this case, the process *misses* many numbers that might be expected to be outputs, setting the stage for logarithms. Students can focus on the sequence formed, its defining characteristics, how to recognize such a sequence, and how to generate the elements either recursively or for any given term using a closed form. Students can explore similar situations looking for generalizations across the situations. For example, they might think about a genealogy question, such as the number of parents or number of grandparents, and its connection to the problem. A parallel development for exponential decay can be made concentrating on the area of the regions obtained by folding paper, leading to a decreasing exponential pattern.

- ***How does this situation promote the habit of looking for generalizations in patterns?***
- ***What are some other ways students might represent their reasoning?***
- ***How can students determine if different forms of representation are equivalent?***
- ***How is the growth rate revealed in the various representations? What does each representation add to the understanding of exponential growth?***
- ***What understandings about functions are essential for developing the concepts of growth and change?***

Comparing Two Patterns of Growth

Students working with situations of exponential growth quickly see that quantities that grow exponentially get very large very rapidly. However, while this is generally an accurate description of exponential growth, it does not capture the essential mathematical qualities. These can best be described in terms of concepts that emerge in the theme of *functions and relations*—dealing with the increment of change of related variables. One way students can begin to appreciate the different kinds of change is to compare two situations, one in which a variable is changing exponentially and the other in which a variable is changing linearly. In the next example, a problem appropriate for middle school students is discussed. The problem is from the setting of *growth and change*—the context is money.

The Gift Problem: Janine's uncle has decided to give her an amount of money every year. He told her that she could have the money in one of two different ways and asked which way she prefers. The two ways he suggests are:

Plan A: She receives $1,000 on her ninth birthday; $1,100 on the birthday after that; $1,200 on the birthday after that, and so forth. That is, she starts with a gift of $1,000 and then the gift she gets increases by $100 every year.

Plan B: She receives $1 on her ninth birthday. (That's right!) Then on the following birthday she receives $2, and on the next it will be $4, and on the next after that it will be $8, and so forth. That is, she starts with a gift of $1 and every year the gift is doubled.

Which would you choose if you were Janine?

Questions to ask: How much will Janine get on her birthdays for the next several years under each of the two plans? Decide which plan you think is best, give your reasons and state any assumptions you made. Does it make any difference how long you think Janine's uncle will give her these gifts?

Suppose that Janine's uncle were to change Plan B so that the first gift was 50¢ instead of $1; would this change your answer? How about 1¢?

Looking into the classroom: Most students begin with the opinion that a pattern that starts with $1, $2, $4, etc., could not possibly ever turn into much money. By computing further values, they begin to notice that, even though the sizes of the gifts are still very, very small for the first 5 or 6 years, the pattern by which they change is somehow different from the "add $100" pattern in Plan A. They are motivated to investigate more values. As the students begin to make tables, they are able to describe the difference in the way the amount changes for each plan. They say things like: "In Plan A, the jumps in the money are always just $100. They never change. But, in Plan B, the jumps keep getting bigger and bigger." They begin to make guesses about how much money Janine would receive "if she got really, really old."

Age	Plan A	Plan B
9	$1,000	$1
10	$1,100	$2
11	$1,200	$4
12	$1,300	$8
13	$1,400	$16
14	$1,500	$32
15	$1,600	$64
16	$1,700	$128
17	$1,800	$256
18	$1,900	$512
19	$2,000	$1,024
20	$2,100	$2,048
21	$2,200	$4,096
22	$2,300	$8,192
23	$2,400	$16,384

The students realize that, once Janine is 21, Plan B leads to larger gifts. However, some notice that the total amount of money she would have at that age is more for Plan A than Plan B. They begin to make a new table that also has columns for calculating the total amount of money she received since she started and find that Plan A is the better one up to age 23.

As the students begin to work on the question of what they would do if the initial gift were 50¢, there is a strong difference of opinion as to whether they need to make all new tables or they can work the problem from the table they already have. One student points out that, if the first gift is 50¢, then the next one will be $1, and the next $2, and so forth. "It will be the same thing except just one year later."

Where Is the Algebra?

This problem involves students in a number of activities that promote algebraic thinking. The table is a kind of symbol system (although not a literal symbol system) in that, once a table has been formed with one set of figures and a set of operations on it mastered by the students, the given input values can be replaced with other sets of input values (such as different size gifts, different constant multiplier in Plan B, or different constant adder in Plan A) and the same reasoning process carried out. They can observe that the Plan B overtakes Plan A 11 years after the first gift. Using an expanded table helps students understand that there are two different functional relationships: the sizes of the gifts and the accumulated gift received by Janine from the beginning to any particular time. As students move to thinking about the general growth patterns, they may use a graph to represent these functions. Students can study the graph of a discrete situation and a continuous situation and discuss what is different and why. If the situations represent continuous functions, students can be asked to estimate the point of intersection of the graphs of the two functions. This encourages students to develop algorithms for making approximations, reasoning from the table or graph to refine their guesses in a systematic way.

Finally, this problem is an occasion to reflect on and analyze two of the most basic types of functional relationships, linear and exponential, and to understand them through an analysis of change as the variable goes from $f(n)$ to $f(n + 1)$. Students can make generalizations and begin to write statements such as $y = 2^n$ or $y = 1000 + (n - 1)100$. In later grades, this problem can be revisited with literal variables in terms of exponential and linear functions, including a table of differences to contrast the rate of change in the two functions. Students should use notation that allows them to compare and abstract: $y = ab^x$ and $y = ax + b$, and recognize how changes in a or b will affect the table and graph and the reverse, how changes in the graph will affect a and b. Students can contrast linear and exponential growth with quadratic and cubic, developing an understanding of the behavior of other functions and of the power of the exponential by considering rate of change as well as other important features.

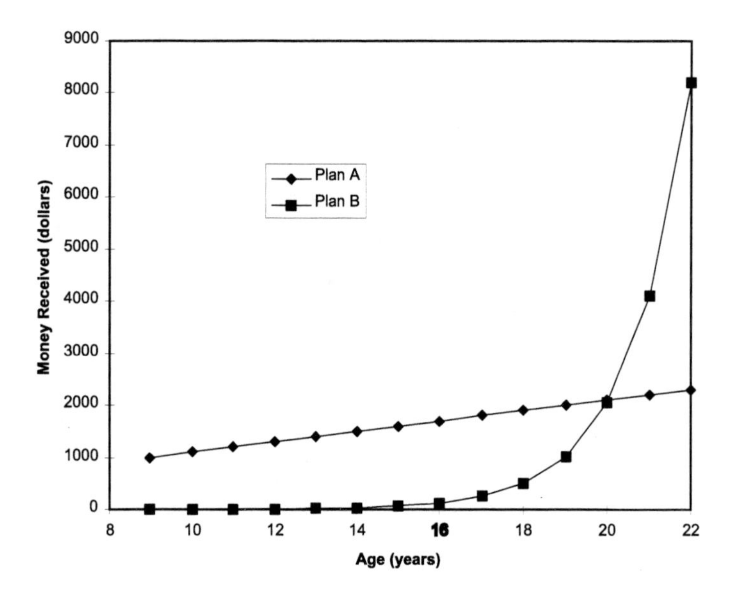

A comparison of Plans A and B

- ***What are some other ways that students might reason about the problem?***
- ***What are some other situations that can be used to compare exponential growth to growth that is not exponential?***
- ***How does understanding one model promote understanding of another model?***

Reversing the Question—Finding the Initial Population in an Exponential Growth Setting

There are many questions in settings of exponential growth that are intrinsically appealing to students and that can be approached either by repeated arithmetic steps or by formulas involving literal variables. One such question is a "backward question" of the standard population growth problem. Ordinarily, we start with a given population, with a given growth rate, and ask what its size will be at some future time. In the reverse version, we start with the size of the population at a later time and ask what the size should be now in order to arrive at the given size. Students who answer such questions by repeated arithmetic are likely to become lost in the problem. At the same time, many students who simply substitute values into a formula are unlikely to ever see the important ideas behind it. However, intermediate notational systems are available for problems of exponential growth that enable students to carry out computations while maintaining an overall grasp of the problem. Such notational systems can also promote a use of literal variables that leads to a use of formulas with genuine understanding.

Summary

The problems have been selected to show how exponential growth can serve as a contextual setting for helping students to develop their algebraic thinking. The problems or a variation can be used from the earliest grades to late high school and illustrate three of the four organizing themes used in the Framework: *functions and relations*, *modeling*, and *language and representation*.

Each of these problems sets out a concrete situation with a puzzling question that invites the student to become engaged with the situation and to solve a genuine problem. As students begin to work on the problem, alone and with others, questions, notational schemes, and suggestions can be provided to help them express, clarify, refine, and reflect on their thinking about the situation and the problem to be solved. Such activities can be important and valuable means for promoting students' understanding of algebra.

Example 2: Contextual Settings Within Size and Shape

Questions about size and shape confront people daily. How big is something? Will this shape or size fit into that place? How do shapes fit together? What happens to other attributes of a shape if one measurement is altered? Answers to these questions are typically thought of as geometric in nature.

This contextual setting demonstrates the interrelationship between geometric and algebraic reasoning, clarifying how algebraic reasoning can be useful to solve problems posed geometrically. For example, problems involving perimeter, area, and volume are situations for thinking about *relationships*. Depending on the grade level, students may be asked to find ways to record relationships symbolically to enhance communication or for use on a computer or calculator. These *representational schemes* can themselves move to the foreground and become the object of discussion and reflection. In some cases, symbolic reasoning provides connections and understandings that might otherwise elude the student. Through discussion, the complementary algebraic and geometric reasoning can evolve. Relationships that appear in a problem can develop and change when the problem is extended and explored in other dimensions.

Reasoning about Patterns

In this activity, designed for the early primary grades in a unit on counting or geometry, students investigate various patterns in a series of squares built from two different colored tiles. Using concrete materials makes the activity appropriate for very young children. The same activity using grid paper instead of tiles is appropriate for older children. The problem focuses on laying the foundations for dimensions, perimeter, and area, just one subset of the landscape of size and shape, as children search for patterns in counting.

The Borders and Blues Problem: The 3 squares were built using white tiles for the border and blue tiles for the interior. Build the squares and count the number of white and blue tiles in each. Can you make the same kind of square using 36 tiles, some white and some blue?

Questions to ask: How many blue tiles are used in each square in this continuing pattern? How many white border tiles are used? Can you make a square with 12 blue tiles?

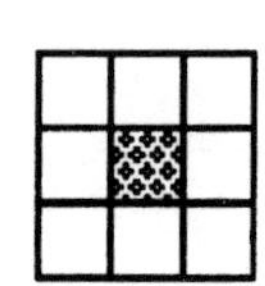
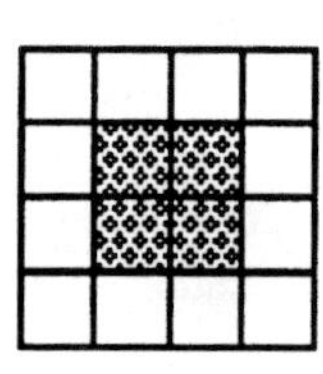
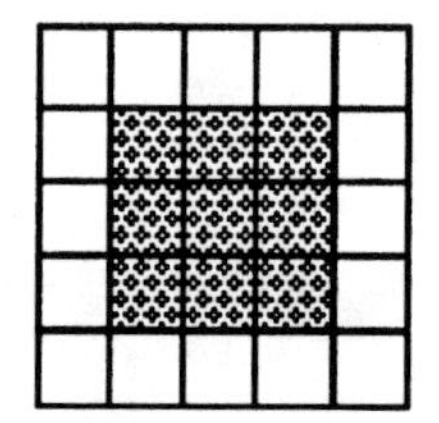

Looking into the classroom: Students build the squares using tiles and record the number of each color. Some forget to label which color is represented by the number they wrote down and need to be reminded. Others draw a picture of the figure and write the numbers next to the picture. To make a square with 36 tiles, some use all white to build the square, then replace the interior tiles with blue. Some build a square with six tiles on the bottom boarder because they recognize the pattern 3, 4, 5, 6, ... Others struggle with different numbers of blue and white tiles, creating rectangles not squares, and are reminded by their classmates about the characteristics of a square.

By shifting the nature of the questions to fractions or to area and perimeter, the problem can be appropriate for upper elementary students. What fraction of the square's area is blue in each figure? What fraction of the area is white? What fraction of the area of the fifth figure in the sequence will be blue? Will the fraction of the area that is blue ever be more than one-half of the area? How does the ratio of border tiles to blue tiles change?

What is the perimeter of each figure? Will the perimeter and the area ever be the same number? Compare the perimeter of the blue portion of the figures to the perimeter of the border.

Connecting Symbols and Situations

Situations involving shape can be good contexts for finding patterns and relationships that connect clearly to symbolic representations. By building or drawing the figures, middle school students can use the way they constructed the shape to reason algebraically about patterns, in a representation that is neither graphical nor symbolic in the traditional algebra sense. Some students reason directly from tables of data from the problem.

The Borders and Blues Problem: The 3 squares have been made using white tiles for the border and blue tiles for the interior. Build the squares and think about how you put the blue and white tiles together. Find a general rule for the number of blue and the number of white tiles for any square.

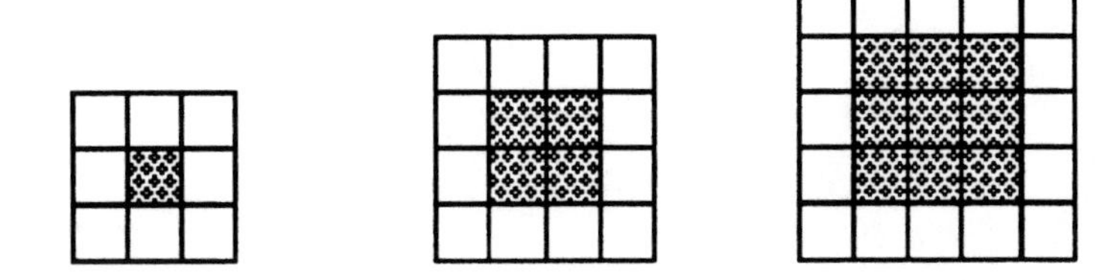

Questions to ask: How can you build the square with six tiles on each edge of the border? Does your rule work for the square with five tiles on a side? How many blue and border tiles will be in the 100th square? How are the formulas related? Will there be a square where the number of white border tiles equals the number of blue tiles? How can the formulas help you answer this question?

In the classroom: Students notice relationships between the number of blue and border tiles as they build. Some see the relation as the number of border tiles is four times the number of blues on a side, plus four (for the four corners).

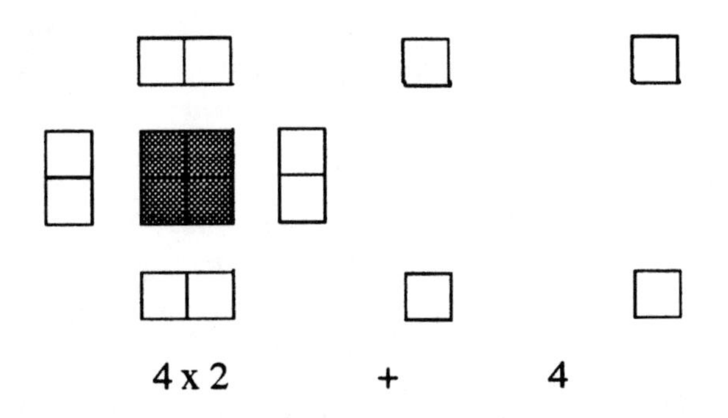

Some students draw a square with six tiles per side and color in the middle: If w is the number of white tiles on the edge of the border, they express their rule as $w^2 - (w - 2)^2$ for the total number of white border tiles; $(w - 2)^2$ for the blue tiles. Other students draw or build the squares differently; their understanding is reflected in the relationships they notice.

Still other students choose to organize their data in a table or a graph. Some notice that the number of border tiles will always be divisible by four. Some students question whether every multiple of four is a border tile total.

Where Is the Algebra?

The focus of this problem at each grade level is on understanding and working with patterns. The theme of representation and language is at the forefront. Explaining precisely, in words, the relationships in this situation becomes cumbersome, and eventually the need to express patterns in symbols leads to a generalization and a coding scheme for the sequence. In the nth blue square, the number of blues is n^2 and the number of white border tiles is $4n + 4$. The total number of tiles used in the nth square is $(n + 2)^2$. This can provide motivation for verifying the identity $(n + 2)^2 = (4n + 4) + n^2$. The students that generate the rule without using the sequence, as the students above did, may use n^2 as the rule for any square and neglect the position in the sequence. Making the correspondence between the first figure and the number of each color, the second figure and the number of each color and so on is an important precursor to working with sequences and series in precalculus. Some students may struggle with notation to express this relationship using symbols such as P1, P2, P3, but from there it is only a short step to more standard notation. Others may need to see this relation clearly laid out in a table.

Figure	Blue tiles	Border tiles	Total tiles
1	1	8	9
2	4	12	16

A discussion of the patterns in the table will bring out recursive thinking: If the number of white border tiles is 12, what happens next?

The theme of function can also form a lens for the problem where, for each figure, students graph (number of blue edge tiles, total number of blue tiles) and (number of blue edge tiles, total number of white tiles) and compare the graphs. Students can reason about the functions to explore when the number of blue tiles overtakes the number of white border tiles; they can find the quotient of two functions to examine the behavior of the ratio of border tiles to blue tiles.

The problem extends into three dimensions, where students can investigate relationships between unit cubes needed to enclose successive central cubes. In the three-dimensional situation, discussions of "border" and "center" engage students in development of clear descriptive language. Students will have diverse ways of thinking about how to find the total number of "border" cubes for a given cube. The patterns in this situation include linear, quadratic, and cubic functions.

- ***What are some other ways that students might reason about the problem?***
- ***Why is it important for students in early grades to reason about patterns?***
- ***Are exercises in symbol manipulation made more meaningful if the symbols are about a familiar concrete situation? How does the presence of a model help students reason algebraically?***
- ***Does the context help bring meaning to the symbolic representation?***

Exploring Complementary Geometric and Algebraic Reasoning

Directly measuring geometric attributes of some shapes is difficult. Indirect measurement, involving use of formulas, is often much simpler; for example, it is easy to compute the area of a triangle given its base and altitude. Developing such formulas involves much more than reasoning about specific measurements of any one exemplar of the shape. Rather, it involves general reasoning about the shape itself. The following problems suitable for middle school or high school demonstrate how algebraic notation is useful to express such reasoning and how symbolic manipulation is useful (needed) to compare various solutions with each other and with the standard formula.

Trapezoid Problem: Explore different ways to find the area of a trapezoid. A diagram of one trapezoid is shown to help you think about the problem. Find ways to express your methods of finding the area of the trapezoid in terms of the height, h, and the lengths of the bases, b_1 and b_2.

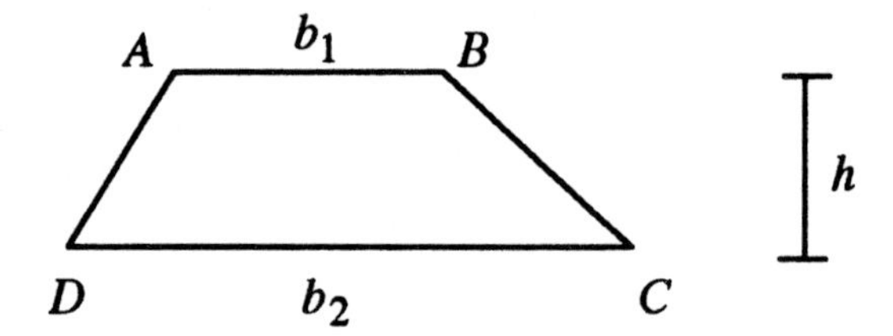

Questions to ask: What different methods did you find? What symbolic expressions did you use to express your reasoning? Are your symbolic expressions equivalent to each other? How can you decide? Can you show that the expressions are (are not) equivalent by symbol manipulation?

Looking into the classroom: Some students use grid paper and subdivide the trapezoid. They find the areas of specific triangles and use these to figure out how they might reason more generally. Other students subdivide the trapezoid into regions in various ways and reason from the diagram. Several of these students create diagrams such as those shown below and develop the corresponding symbolic expressions.

Students who are facile with symbol manipulation attempt to show that their symbolic expressions are equivalent. Other students who are less facile with symbol manipulation use measurements from specific examples to verify the equivalence of expressions. Still others use a symbol manipulator to demonstrate equivalence or dynamic geometry software to investigate the situation.

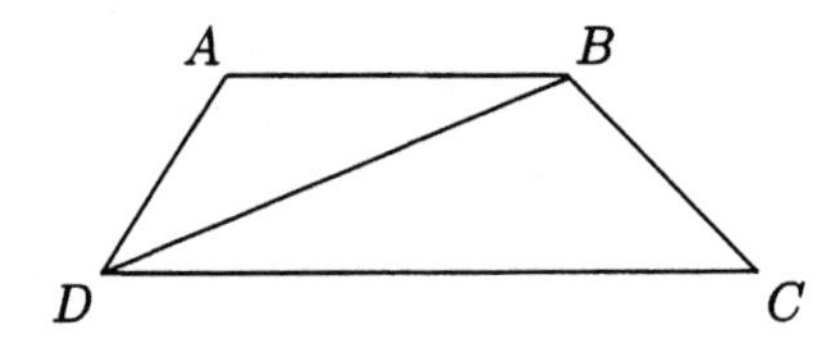

Area of trapezoid *ABCD* = area of triangle *ABD* + area of triangle *BCD.*

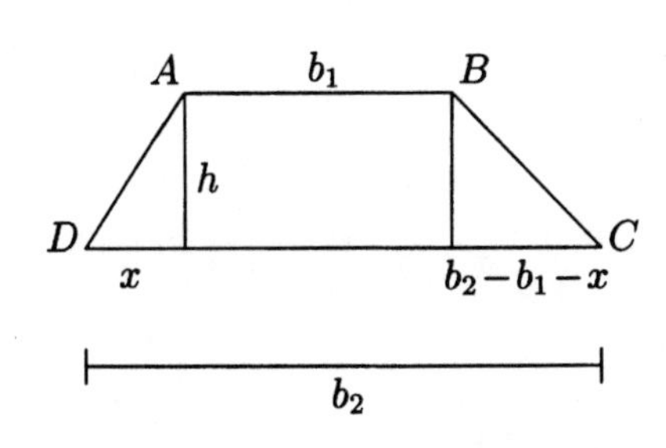

Area of trapezoid *ABCD* = area of left triangle + area of rectangle + area of right triangle.

$$= \frac{1}{2}hx + b_1h + \frac{1}{2}h(b_2 - b_1 - x)$$

$$= \frac{1}{2}hx + b_1h + \frac{1}{2}hb_2 - \frac{1}{2}hb_1 - \frac{1}{2}hx$$

$$= \frac{1}{2}hb_1 + \frac{1}{2}hb_2$$

Where Is the Algebra?

Partitioning a trapezoid into subregions and writing symbolic expressions for those regions demonstrates how algebraic notation can be used to express geometric reasoning. The process of expressing a geometric idea using algebraic symbols and transforming the symbols into a new statement that has a different geometric interpretation gives added meaning to manipulation of algebraic symbols. The symbols describe the geometric reasoning that directs the process. Here, the power of algebra is in the transition between reasoning with symbols and interpreting in the context of the problem. Questions of proof and what constitutes proof arise when equivalence of expressions is verified numerically for specific trapezoids.

The **Trapezoid Problem** can be extended when students are familiar with similarity and can use proportional reasoning.

- ***How does use of a symbol manipulator impact students' learning to reason algebraically? What do students need to understand about symbols and symbol manipulation to use a symbol manipulator? What level of ability to manipulate symbols is prerequisite to this understanding?***
- ***What are some other ways to think about the area of a trapezoid? How do the symbolic and geometric representations model the way students reasoned about the area of a trapezoid?***
- ***When are students able to appreciate the need for proof? At what stage is formal proof appropriate or necessary? How do they learn what constitutes an acceptable proof? Can problems like these help students understand and appreciate the power and elegance of algebra?***

Sam's Problem: To solve the Trapezoid Problem, Sam extended the lines *AD* and *BC* to form triangle *DPC*. He found the area of the trapezoid by subtracting the area of triangle *APB* from the area of triangle *DPC*. Sam used the similarity of the two triangles to express the difference in their areas only in terms of the height, h, of trapezoid *ABCD* and the lengths of its bases, b_1 and b_2. Will Sam's method lead to the standard form for the area of a trapezoid?

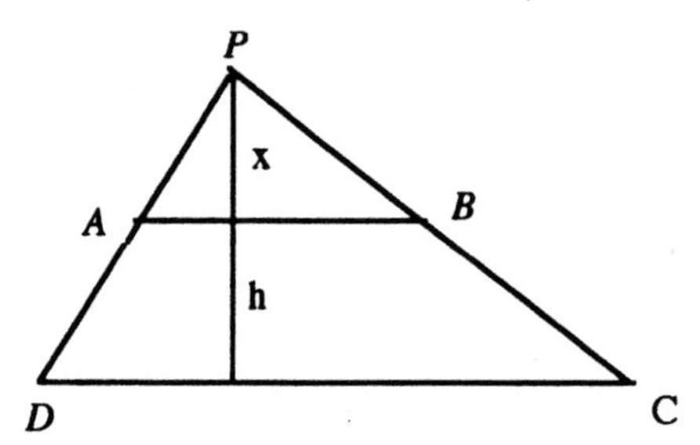

Looking into the classroom: Some students will use symbols to express their thinking and reasoning as follows:

area of trapezoid $ABCD = \frac{1}{2}(x+h)b_2 - \frac{1}{2}xb_1$

Similarity of the triangles gives $x = \frac{b_1 h}{b_2 - b_1}$

Therefore, area of trapezoid $ABCD = \frac{1}{2}\left(\frac{b_1 h}{b_2 - b_1} + h\right)b_2 - \frac{1}{2}\left(\frac{b_1 h}{b_2 - b_1}\right)b_1$

Where the student goes from here depends on availability of a symbol manipulator or facility with symbolic manipulation. Using the distributive property, the area of trapezoid *ABCD* can be written as the equivalent expression, $\frac{1}{2}h(b_1 + b_2)$. This expression leads some students to wonder if there is another way to cut and rearrange the trapezoid to get a triangle whose base is $(b_1 + b_2)$ and height is h.

Where Is the Algebra?

Symbolic manipulation can be used to simplify complicated expressions using the properties of real numbers. With experience, students learn both when to attempt to simplify an expression and how to tell if what they get is, in some sense, simpler. Students who observe that the expression for the area of a trapezoid can be written as $\frac{1}{2}h(b_1 + b_2)$ and who wonder if there is a different way to find the area are at a different level of symbolic reasoning. They are using equivalent expressions to look for other patterns in the problem.

Both the **Trapezoid Problem** and **Sam's Problem** can be extended to three dimensions for high school students. For example, students might explore different ways to find the volume of the frustum of a pyramid using Cavalieri's Principle and the fact that the volume is one-third the area of the base times the height, expressing their methods symbolically in terms of the dimensions of the frustum. They can generalize Sam's method to three dimensions to find the volume of the frustum. Reasoning in three dimensions is challenging for many students. They can reflect on the methods they developed in two dimensions and use them as the basis for reasoning in three dimensions. Extending results to higher dimensions can lead to reasoning about what is involved in generalizing a result. Algebraic notation can help direct the reasoning.

Summary

The problems in this example are obviously geometric in character, but are embedded within the theme of *representation.* Assigning variables to represent carefully chosen measurements can clarify relationships found in geometric shapes. An early example of this may be when students realize that the length and width of a rectangle always control the area. Being able to generalize this relationship in a formula allows students to work with the mathematical model to develop further relationships, from the simple $A = l \times w$ to the closely related $A = \frac{1}{2}\, b \times h$, to the various forms of the area of a trapezoid and other shapes. In some problems, such as the **Border Problem**, students connect numerical patterns to an underlying geometry to develop a representation.

In addition to working with algebraic *models* to investigate shapes, students can also explore relationships among the zero-, one-, two-, and three-dimensional characteristics of the same geometric shape. Using tables and graphs to represent the problem, and finding patterns and relationships within and reasoning from tables and graphs, facilitates such investigations. Students may be searching for relationships across one line of a table, or they may be using recursive patterns. Symbolic expressions not only represent relationships among attributes of a geometric shape, but also, along with tables and graphs, provide efficient ways of exploring, describing, and verifying patterns. This is particularly useful when students track and record sophisticated ideas. (What results do changes in linear dimensions of a shape cause in two- and three-dimensional measurements? Can you predict the elements of the formula for the frustum of a pyramid from your experience with Sam's method for a trapezoid?) Using all three methods of representing problems in the size-and-shape setting raises the question of which is more effective for a particular problem or if other methods (visualization, for example) are preferred. In many cases, the ability to represent a problem in several ways may lead to complementary insights.

The problems given above are intended as the beginning point for further questions about where algebra can be found and what purpose it serves in the setting of size and shape. Algebra was used as a way to describe *relationships* and to represent problems. Algebraic reasoning was used to find equivalent solutions or to verify observations. Underlying all the problems is also the theme of *structure,* how algebra complements geometric reasoning. Concrete objects, pictures, diagrams, and symbols were some of the ways used to *represent* the situations. While this example was posed in the setting of *size and shape,* it also intersected with ideas from *growth and change* and *number.*

Example 3: Contextual Settings Within Number

Another setting people experience is that of number. People use number to order and categorize their world. Through number, they carry out the computations necessary for their work and their daily routine. Children's understanding of even/odd allows them to make judgments about sharing, choosing teams, and taking turns. As children discover that they can count forever, they develop a sense of magnitude.

The study of numbers includes many important topics (structure of numbers, number patterns, operations, relations, properties of numbers, and methods of counting, ordering, and computing) that underlie different organizing themes. Number is essential to understanding the behavior of *functions* and *relations.* The order and characteristics of the real number system often provide the first example children have of mathematical *structure.*

The search for generalizations and methods of *representation* have many origins in the setting of number. As students reason about calculations, they may observe numerical patterns or represent quantities pictorially or geometrically; with advances in technology, reasoning about number has broadened to include use of spreadsheets and sets of instructions for computers. Ultimately, reasoning about calculations produces a need for symbolic notation, both for representing the process and for modeling relations. The focus of instruction should not, however, be on the manipulation of symbols but rather on the conceptual understanding of the meaning of the symbolic representation. Choosing number as a setting provides opportunities to illustrate how student understandings of any of the themes can be developed singly or in conjunction with others.

The properties of the real number system allow us to rewrite expressions in an equivalent form, to represent a situation, and then to work within that representation free from context. A focus on the distributive property as one illustration of how number is incorporated into the framework highlights the *structural* aspect of algebra. Other properties of number could serve the same purpose. Other organizing themes, however, are also present as students develop and apply their understanding.

Understanding the distributive property allows students to

- think about different characteristics of the same situation;
- represent, generalize, and confirm conjectures;
- gain new information about a situation from an equivalent representation.

With the distributive property, students can think about the calculation embedded in a situation as a sum of quantities, each of which has two or more factors, $ab + ac$, or they can think of the same situation as a product of quantities, each of which may be a sum of two or more addends, $a(b + c)$. Generalizing and gaining new information from a representation are equally possible for many of the properties of number.

The following problem situations illustrate these different aspects of the distributive property. However, because these outcomes are not disjoint, the problems are not separated according to specific goals. Instead, the problems are grouped according to developmental level of students. They are intended to serve as a springboard for *thinking* about the role of the distributive property in reasoning algebraically and for further discussion of the experiences necessary for students as they construct a view of algebra from their work with number.

Laying the Foundation

Early childhood experiences that help children understand the setting of number include investigations into patterns and regularities. Reasoning about the relationship between quantities and about efficient ways to link two mathematical representations involves using properties such as the distributive property as a matter of course. Using situations that have two equivalent interpretations will help build understanding of the distributive property. Experiences with the distributive property can occur in many situations in the elementary grades: simplifying computation $3 \times 10 + 3 \times 9 = 3(10 + 9)$ or using an area model to show how to express multiplication of two-digit numbers.

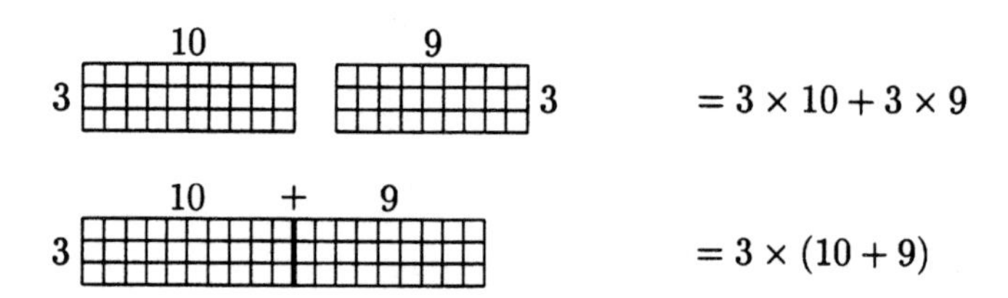

A visual representation allows students to assimilate the equivalence of the two ways to think.

- ***Why is it important for students to have alternative ways to calculate?***

- ***What does this area model add to student understanding of the multiplication algorithm for whole, fractional, and decimal numbers?***

Building on the Foundation

As students in the middle grades become more familiar with expressing relationships symbolically, individual students may create different but equivalent expressions, depending on how they reasoned about the problem. The following example illustrates how the distributive property provides a link between different ways of thinking.

The Telephone Network Problem: There are many different houses in a particular region. How many different telephone paths are necessary if each house is directly connected to every other house?

Looking into the classroom: Ann and Juan decide to begin with some small examples and see if they can find a pattern. Ann begins with 4 and then 5 houses and draws the number of connections; Juan uses 6 and then 7. They make a table and discover they can tell how many connections they need for 8 houses, then 9, then 10. They can see a way to get to the NEXT entry in the table, but they cannot see how to get any general rule.

Number of Houses	Number of Paths
4	6
5	10
6	15
7	21
8	**NEXT = previous Number of Houses + previous Number of Paths**

Another student writes, "If there are N houses, you won't connect a house to itself,

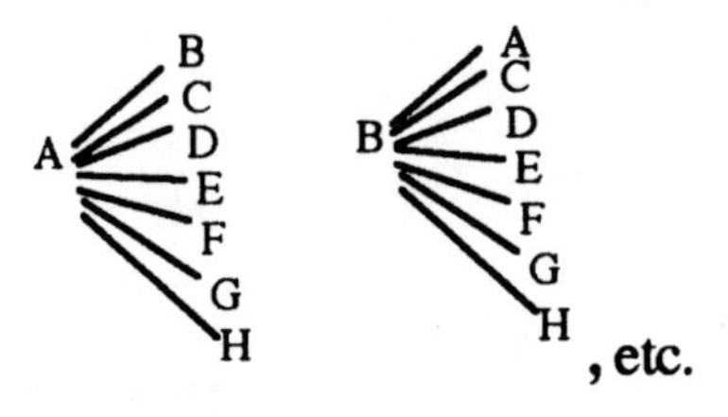

so each will have to be connected to one less or $n - 1$ houses. This means there will be $n(n - 1)$ connections. For example, for 8 houses...

If house A is connected to house B, though, it is the same as if house B is connected to house A, so my answer should be divided by 2. There will be $n(n - 1)/2$ connections."

Another pair of students makes a chart where each "1" represents a connection between the houses and a "0" represents no connection.

	A	B	C	D . . . n
A	0	1	1	1 . . . 1
B	1	0	1	1 . . . 1
C	1	1	0	1 . . . 1
⋮				
n	1	1	1	1 . . . 0

0 = no connection
1 = connection

The chart is a square, so if there are n houses, there will be n^2 connections in the chart.

But the diagonal is not helpful because that shows a house is connected to itself. There are n houses on the diagonal, so the number of connections is $n^2 - n$. The connections on each half of the diagonal are the same (A to B is the same as B to A), so really there are only $(n^2 - n)/2$ connections. Tina rewrote the rule as

$$1 \times n \times (n-1)/2$$

She feels sure that this is equivalent because she recognizes she can apply the distributive property to $n^2 - n$. She comments that the formula for the area of a triangle is $\frac{1}{2}bh$ and wonders if this relates to the matrix she and her partner made.

Questions to ask: Do all of the rules generate the same answers as Ann's NEXT rule? Are the rules equivalent to each other? How do you know? One of the rules looks like the formula for the area of a triangle, but where is the triangle? Would a graph facilitate understanding of different reasoning processes?

Where Is the Algebra?

Searching for recursive patterns can provide a solution but is not efficient in many cases. Relations are often difficult to describe in closed form. Those who try to find a more direct rule may think very differently and use different symbolic representations. One link that shows these different ways of thinking are mathematically equivalent is the distributive property. Tina's contribution is somewhat different in nature. She has deliberately applied the distributive property and linked the form of the result to something else she knows. Students can try to connect her new representation to something concrete in the problem. Some representations of this problem will have a triangular aspect because the numbers generated are the triangular numbers. Some students may recognize these from Pascal's triangle. Students may reason from the sequence of numbers or the ordered pairs, as follows,

Number of houses	Number of paths	Ordered pairs
1	0	(1,0)
2	1	(2,1)
3	3	(3,3)
4	6	(4,6)

and think about what relation operating on the first would give the second. Others may reason from a graph of the ordered pairs. They should see that the resulting graph is like the graph of $y = x^2$ but with a smaller rate of change and different vertex. This is an occasion to discuss which of the two representations, $y = x^2 - x$ or $y = x(x - 1)$, gives the most insight on estimating answers. For very large numbers, the graphs of $y = x^2$ and $y = x^2 - x$ are almost identical.

- ***How do the representations capture the reasoning process of the students? What is the advantage of having different representations?***
- ***How are the suggestions qualitatively different from each other? How can you systematically build on such thinking?***

At the high school level, the problem can be extended to a counting problem where there are n ways to make the first choice, $n - 1$ ways to make the second choice, and so $n(n - 1)$ ways to choose two things. Dividing by 2 reduces the repetition. The triangular numbers that result, 1, 3,

6, 10, ... , can be related to the binomial theorem, $(x + y)^n$ as the coefficients of the terms in expanded form, a powerful application of the distributive property. When students learned to multiply two-digit numbers, a geometric model of the distributive property can provide them with an understanding of the algorithm. For example, a student might write 3(27) = 3(20 + 7) = 60 + 21 = 81. When the computational process is extended to 27(42), the distributive property leads to (20 + 7)(40 + 2), or the product of two binomials. Initially, some students may say symbolically that $(x + y)^2 = x^2 + y^2$ but drawing an area model can convince them the square $(x + y)$ has area $x^2 + 2xy + y^2$ and reinforces use of the distributive property.

	x +	y
x +	x^2	xy
y	yx	y^2

The geometric model reinforces the need for the cross products xy. The use of $(x + y)$ as a factor being distributed as an entity, $(x + y)x + (x + y)y$, is an extension of the thinking established in the early grades and helps explain symbolically why the sum of the two squares as an answer to $(x + y)^2$ is insufficient. Students should be able to think and reason about the process and the algorithm they use. Expanding a trinomial or cube enables students to think of reasonable ways to extend their geometric model both in two and three dimensions and allows them to begin to develop patterns that will later lead to the binomial expansion theorem

$$(x + y)^n = x^n + nx^{n-1}y + \ldots + y^n.$$

Physical models can help students understand a situation, but students should become more and more comfortable with symbols as ways to represent situations, as well as ways to generalize arithmetic. The follow-up questions deliberately focus on *structural* issues: When does this property apply? What is the pattern in the terms? The comparative usefulness of the geometric model or the symbolic representation highlights choices of *representations.*

- ***How much emphasis should be placed on structure at each level? What essential understandings do students need? How are these to be developed in contexts that motivate?***
- ***How much exposure and over what length of time does it take for students to recognize and be able to use the distributive property?***

Within the setting of number and the organizing theme of structure, it becomes important to reflect on certain patterns and behaviors in very general ways. Some situations seem to share certain characteristics. These common attributes allow you to generalize and treat in similar ways systems that, on the surface, seem to have little in common. The distributive property is actually a specific instance of a more general structural property. College- bound students might investigate the following:

The Function Problem: Think about the functions you have studied. Which of these functions have the property $f(a + b) = f(a) + f(b)$?

Looking into the classroom: Students check the functions they recall. For example they write "Is $\sin(a + b) = \sin a + \sin b$?" Some check the validity of each conjecture by making substitutions. Others discuss the nature of the function: "The value of $\sin x$ is never greater than 1, but sin a + sin b could be." The students continue to investigate other functions.

Questions to ask: What do the functions you chose have in common? Would the property, $f(a * b) = f(a) * f(b)$, apply to operations other than adding for the functions you identified?

Where Is the Algebra?

Students may conclude that $f(a + b) = f(a) + f(b)$ applies only to linear functions such as those of the form $y = kx$. Some students find a geometric argument convincing:

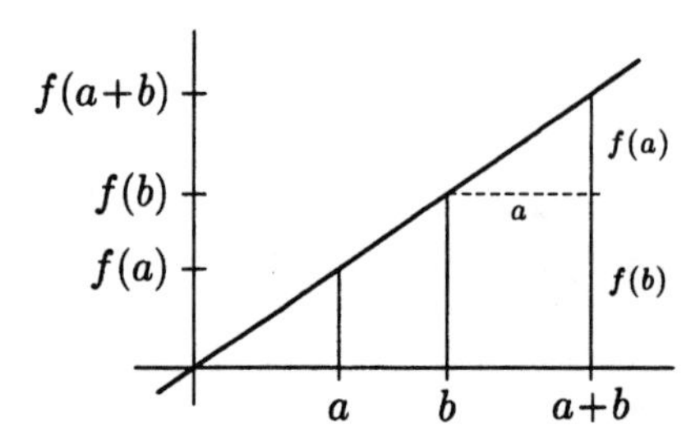

Those who have studied abstract algebra will recognize the general question: what operations are preserved by a mapping from one algebraic structure to another? For the linear function, $y = kx$ with k fixed, each real number x is mapped onto the real number kx. Understanding what is preserved under what conditions allows flexibility in analyzing situations; in some cases, it is easier to analyze the range elements after the transformation, in others, to begin with the original domain elements then make the transformation. This study of what is preserved under different mappings leads to the concept of homomorphic functions, those for which $f(a * b) = f(a) * f(b)$. The distributive property of multiplication over addition is a particular example of a homomorphic mapping.

What are the connections between the distributive property of multiplication over addition and linearity?

Summary

While the setting for this example is number, the theme of structure comes to the forefront. The distributive property also plays a role in operating on and understanding relationships in systems where symbols represent objects other than number. Scalar multiplication distributes over addition of matrices; if the dimensions are aligned, matrix multiplication distributes over matrix addition.

To use the structural properties of an algebraic system, students have to be introduced to these properties in thoughtful and meaningful ways in the early grades, using problem settings that are familiar to both students and teachers. The ability to generalize relationships is at the heart of algebra, and the structural nature of the system allows relationships to be expressed in different forms, some more useful in particular situations than others. Each different representation can add to the knowledge gained from one of the other representations. The use of symbols and the properties governing the behavior of operations with those symbols allow students to represent a situation in symbols, to manipulate those symbols temporarily free from the situational meaning to gain insights and information about the situation, and then to return to the situation to make sense of the symbols.

- ***What is the interface between technology and structure?***
- ***How much symbol manipulation is necessary to function effectively with symbolic representations?***

USING THE FRAMEWORK

Groups from the mathematics education community can use the ideas in this document to organize their thinking and discussion of school algebra to develop a vision of algebra in a K-12 curriculum and as a means to move toward that vision. Possible questions might include the following:

- What is the essential nature of each theme?
- How do themes help organize ideas?
- Are some themes more appropriate at different grade levels than at others?
- Is there a hierarchy to the themes?
- What contextual settings can be fruitful grounds for exploring algebraic concepts?
- How do the themes focus algebra in grades 9-12?
- What are examples of curricula that provide a coherent and balanced algebra curriculum in grades K-12?
- How can adequate articulation and continuity be built into a K-12 algebraic sequence?
- What should be done to help students develop depth in their algebraic understanding?
- How and when should algebraic understanding be assessed?
- What are some characteristics of algebraic reasoning at different developmental levels?

The demands of a fast-changing society and the presence of technology require a vision of school algebra that is dynamic and fluid enough to keep pace with future needs of society, yet retains the essential aspects of algebra that have made it so significant in the history of mathematics. The workforce needs citizens who can adapt to new technologies, identify problems, reason about problems, and communicate their findings using symbols, graphs, tables, pictures, and words. Studies from other countries demonstrate that students can learn to reason algebraically much earlier than grade 9 and that all students can do so. Developing and implementing a coherent and balanced algebra curriculum for grades K-12 requires a complete rethinking of the entire mathematics curriculum, a task that is already underway by some involved in curriculum development.

The success of implementing any vision of school algebra ultimately lies in creating conditions, policies, assessment, curriculum materials, and support that enable teachers to provide the kind of algebra experience that is essential for all students. Discussions and policies on "who takes algebra when" must have the full participation of teachers who will be responsible for enacting the changes. The algebra that is called for in this document is quite different from the algebra that most teachers have been taught or have been teaching. A framework such as this one can support teachers in a critical and reasoned review and adaptation of new curriculum materials purporting to exemplify a rethinking of algebra. Long-term professional development activities and preservice programs must be examined in light of a framework and the issues raised while equal efforts must be made to help "lay people" understand the reasons for and nature of a different view of algebra. Reconfiguring algebra as a K-12 endeavor will take time, commitment, and deep thinking on the part of the entire mathematics education community to make a successful algebraic experience a reality for all students.

BIBLIOGRAPHY

This bibliography will be a useful resource for anyone interested in the nature and role of algebra in the K-12 curriculum.

Arcavi, A. (1994). "Symbol sense: Informal sense-making in formal mathematics." *For the Learning of Mathematics,* 14 (3), 24-35.

Bastable, V., & Schifter, D. (1997). "Classroom stories: Examples of elementary students engaged in early algebra," in J. Kaput (Ed.), *Employing Children's Natural Powers to Build Algebraic Reasoning in the Content of Elementary Mathematics.*

Beaton, A.E., Mullis, I.V.S., Martin, M.O., Gonzalez, E.J., Kelly, D.L., & Smith, T.A. (1996). *Mathematics Achievement in the Middle School Years: IEA's Third International Mathematics and Science Study.* Chestnut Hill, MA: Center for the Study of Testing, Evaluation, and Educational Policy, Boston College.

Birkhoff, G. (1973). " Current trends in algebra." *American Mathematical Monthly,* 80, 760-782.

Birkhoff, G., & MacLane, S. (1992). " A survey of modern algebra: The fiftieth anniversary of its publication." *Mathematical Intelligencer,* 14 (1), 26-31.

Blais, D.M. (1988). "Constructivism—A theoretical revolution for algebra." *Mathematics Teacher,* 81, 624-631.

Burrill, B., & Burrill, J.C. (1992). *Data Analysis and Statistics Addenda Series, Grades 9-12.* Reston, VA: NCTM.

Carlson, D., Johnson, C., Lay, D., & Porter, A. (1993). "The linear algebra curriculum study group recommendations for the first course in linear algebra." *The College Mathematics Journal,* 24, 4-46.

Chambers, D.L. (1994). "The right algebra for all." *Educational Leadership,* 51 (6), 85-86.

Cipra, B. (1988). "Recent innovations in calculus instruction, " in L. Steen (Ed.)., *Calculus for a New Century* (pp. 95-103). Washington, DC: Mathematical Association of America.

Cobb, P., Wood, T., & Yackel, E. (in press). "Learning through problem solving: A constructive approach to second grade mathematics," in E. von Glasersfeld (Ed.), *Constructivism in Mathematics Education.* Dordrecht, The Netherlands: Reidel.

Coburn, T.G. (1993). *Patterns Addenda Series.* Reston, VA: NCTM.

Conference Board of the Mathematical Sciences (1983). *The Mathematical Sciences Curriculum K-12: What Is Still Fundamental and What Is Not.* Report to the NSB Commission on Precollege Education in Mathematics, Science, and Technology. Washington, DC: Author.

Confrey, J. (1995). "Student voice in examining splitting as an approach to ratio, proportions, and fractions," in L. Miera & D. Carraher (Eds.), *Proceedings of the 19th Annual Conference for the Psychology of Mathematics Education, Vol. 1,* pp. 3-29. Recife, Brazil.

Confrey, J., & Smith, E. (1995). "Splitting, covariation, and their role in the development of exponential functions." *Journal for Research in Mathematics Education,* 26, 66-86.

Confrey, J. (1994). "Splitting, similarity, and rate of change: A new approach to multiplication and exponential functions," in G. Harel & J. Confrey (Eds.), *The Development of Multiplicative Reasoning in the Learning of Mathematics.* Albany: SUNY Press.

Coxford, A.F., & Shulte, A.P. (Eds.) (1988). *Ideas of Algebra, K-12.* Reston, VA: NCTM.

Cuoco, A., & LaCampagne, C.B. (submitted to the *Notices* of the AMS.) "Department of Education launches algebra initiative."

Cuoco, A. (in press). "Early algebra and structure of calculations, " in J. Kaput (Ed.), *Employing Children's Natural Powers to Build Algebraic Reasoning in the Content of Elementary Mathematics.*

Cuoco, A. (1993). "Action to process: Constructing functions from algebra word problems." *Intelligent Tutoring Media*, 4 (3/4), 118-127.

Day, R.P. (1993). "Solution revolution." *Mathematics Teacher*, 86 (1), 15-22.

Edwards, E.L. (Ed.) (1990). *Algebra for Everyone.* Reston, VA: NCTM

Eisenberg, T. (1992). " On the development of a sense for functions," in G. Harel & E. Dubinsky (Eds.), *The Concept of Function: Aspects of Epistemology and Pedagogy.* (*MAA Notes*, Vol. 25, pp. 153-174). Washington, DC: Mathematical Association of America.

Euler, Leonard (1984). *Elements of Algebra.* (Translated by John Hewlett.) NY: Springer-Verlag.

Ferrini-Mundy, J. & Johnson, L. (1994). "Recognizing and recording reform in mathematics: New questions, many answers." *Mathematics Teacher*, 87 (3), 190-193.

Fey, J.T., Fitzgerald, W.M., Friel, S.N., Lappan, G.T., & Phillips. E.D. (1994). *Bits and Pieces, Part I.* Connected Mathematics Project. (Limited circulation pilot edition.) East Lansing, MI.

Fey, J.T. (1989). "School algebra for the year 2000," in S. Wagner & C. Kiernan (Eds.), *Research Issues in the Learning and Teaching of Algebra* (pp. 199-213). Reston, VA: NCTM. Hillsdale, NJ: Erlbaum.

Fey, J.T., & Good, R. (1985). "Rethinking the sequence and priorities of high-school mathematics curricula." In C. Hirsch (Ed.), *The Secondary School Mathematics Curriculum.* 1985 Yearbook of the National Council of Teachers of Mathematics. Reston, VA: NCTM.

Freudenthal. H. (1983). *Didactical Phenomenology of Mathematical Structures.* Dordrecht, The Netherlands: Reidel.

Goldenberg, E.P. (1988). "Mathematics, metaphors, and human factors: Mathematical, technical, and pedagogical challenges in the educational use of graphical representation of functions." *Journal of Mathematical Behavior*, 7, 135-173.

Grouws, D.A. (Ed.) (1992). *Handbook of Research on Mathematics Teaching and Learning.* Reston, VA: NCTM. New York: Macmillan.

Hawkins, B.D. (1993). "Math: The great equalizer—Equity 2000 and QUASAR, improving minority standing in gatekeeper courses." *Black Issues in Higher Education*, 10 (6), 38-41.

Harel, G., & Confrey, J. (Eds.) (1994). *The Development of Multiplicative Reasoning in the Learning of Mathematics.* Albany: SUNY Press.

Heid, M.K. (Ed.) (1995). *Algebra in a Technological World Addenda Series, Grades 9-12.* Reston, VA: NCTM.

Heid, M.K. (1988). *The Impact of Computing on School Algebra: Two Case Studies Using Graphical, Numerical, and Symbolic Tools.* Proceedings of ICME-6, Theme Group 2, Working Group 2.3. Budapest, Hungary.

Hiebert, J., & Wearne, D. (1991). "Methodologies for studying learning to inform teaching," in E. Fennema, T.P. Carpenter, & S.J. Lamon (Eds.), *Integrating Research on Teaching and Learning Mathematics* (pp. 153-176). Albany: SUNY Press.

Johnston, W.B., & Packers, A.E. (1987). *Work Force 2000: Work and Workers for the Twenty-First Century.* Indianapolis: Hudson Institute.

Kaput, J. (Ed.) (in preparation). *Employing Children's Natural Powers to Build Algebraic Reasoning in the Context of Elementary Mathematics.*

Kaput, J. (forthcoming). *Integrating Research on the Graphical Representation of Functions.* Hillsdale, NJ: Erlbaum.

Kaput, J. (in press). "Democratizing access to calculus: New routes using old roots," in A. Schoenfeld (Ed.), *Mathematical Thinking and Problem Solving.* Hillsdale, NJ: Erlbaum.

Kaput, J. (1987). "Representation systems in mathematics," in C. Janvier (Ed.), *Problems of Representation in the Teaching and Learning of Mathematics* (pp. 19-26). Hillsdale, NJ: Erlbaum.

Karpinski, L.C. (1917). "Algebraical development among the Egyptians and Babylonians." *American Mathematical Monthly*, 257-265.

Katz, V. (1995). "The development of algebra and algebra education," in C. LaCampagne, W. Blair, & J. Kaput (Eds.), *The Algebra Initiative Colloquium (Vol. 1).* Washington, DC: U.S. Department of Education.

Kieran, C. (1994). "A functional approach to the introduction of algebra—Some pros and cons," in J.P. da Porte & J.F. Matos (Eds.), *Proceedings of the Eighteenth International Conference on the Psychology of Mathematics Education*, 1, 157-175. Lisbon, Portugal.

Kieran, C. (1992). "The learning and teaching of school algebra," in D.A. Grouws (Ed.), *Handbook of Research on Mathematics Teaching and Learning* (pp. 390-419). New York: Macmillan.

Kieran, C. (1989). "The early learning of algebra: A structural perspective," in S. Wagner & C. Kieran (Eds.), *Research Issues in the Learning and Teaching of Algebra* (pp. 33-56). Reston, VA: NCTM. Hillsdale, NJ: Erlbaum.

Kleiner, I. (1989). "Evolution of the function concept: A brief survey." *College Mathematics Journal*, 20 (4), 282-300.

Kline, M. (1972). *Mathematical Thought from Ancient to Modern Times.* New York: Oxford University Press.

Leinhardt, G., Zaslavsky, O., & Stein, M.K. (1990). "Functions, graphs, and graphing: Tasks, learning, and teaching." *Review of Educational Research*, 60 (1), 1-64.

Leitzel, J.R.C. (Ed.) (1991). *A Call for Change. Recommendations for the Mathematical Preparation of Teachers of Mathematics.* Washington, DC: Mathematical Association of America.

Leiva, M.A. (Ed.) (1991). *Curriculum and Evaluation Standards for School Mathematics Addenda Series, Grades K-6.* Reston, VA: NCTM.

Michigan Council of Teachers of Mathematics (1990). *Algebra Activities, K-9.* Lansing, MI: Author.

Moses, B. (1993). "Algebra, the new civil right." Paper presented at the SUMMAC II Conference. Cambridge, MA.

National Commission on Excellence in Education (1983). *A Nation at Risk: The Imperative for Educational Reform.* Washington, DC: U.S. Government Printing Office.

National Council of Teachers of Mathematics (1994). "Board Approves Statement on Algebra." *NCTM News Bulletin.*

National Council of Teachers of Mathematics (1992). "Algebra for the Twenty-First Century." *Proceedings of the August 1992 NCTM Conference.* Groton, MA.

National Council of Teachers of Mathematics (1991). *Professional Standards for Teaching Mathematics.* Reston, VA: Author.

National Council of Teachers of Mathematics (1990). E.L. Edwards, Jr. (Ed.), *Algebra for Everyone.* Reston, VA: Author.

National Council of Teachers of Mathematics (1989). *Curriculum and Evaluation Standards for School Mathematics.* Reston, VA: Author.

National Council of Teachers of Mathematics (1980). *An Agenda for Action: Recommendations for School Mathematics of the 1980s.* Reston, VA: Author

National Research Council (1990). Steen, L.A. (Ed.), *On the Shoulders of Giants: New Approaches to Numeracy.* Washington, DC: National Academy Press.

National Research Council (1989). *Everybody Counts: A Report to the Nation on the Future of Mathematics.* Washington, DC: National Academy Press.

National Science Board Commission on Precollege Education in Mathematics, Science, and Technology (1983). *Educating Americans for the Twenty-First Century: A Plan of Action for Improving Mathematics, Science, and Technology Education for All American Elementary and Secondary Students So That Their Achievement Is the Best in the World by 1995.* Washington, DC: National Science Foundation.

Nesher, P. (1986). "Are mathematical understanding and algorithmic performance related?" *For the Learning of Mathematics,* 2-9.

Pelavin, S., & Kane, M. (1988). *Minority Participation in Higher Education.* Prepared for the U.S. Department of Education.

Phillips, E. A., Gardella, T., Reely, C., & Steward, J. (1991). *Patterns and Functions Addenda Series, Grades 5-8.* Reston, VA: NCTM.

Rachlin, S. (1982). "Processes used by college students in understanding basic algebra." Columbus, OH: ERIC Clearinghouse for Science, Mathematics, and Environmental Education (SE 036 097).

Reese, C.M., Miller, K.E., Mazzeo, J., & Dossey, J.A. (1997). *NAEP 1996 Mathematics Report Card for the Nation and the States.* Washington, DC: National Center for Education Statistics.

Reys, R.E., & Nohda, N. (1994). *Computational Alternatives for the Twenty-First Century: Cross-Cultural Perspectives from Japan and the United States.* Reston, VA: NCTM.

Russell, S., et al. (1995). *Investigations of 3rd/4th Grade Interpreting Graph Units.* Palo Alto, CA: Dale Seymour Publications.

Schifter, D. (Ed.) (in press). *Voicing the New Pedagogy: Teacher Narratives and the Construction of Meaning for the Rhetoric of Mathematics Education Reform.* New York: Teachers College Press.

Schoenfeld, A.H. (1985). "Metacognitive and epistemological issues in mathematical understanding," in E.A. Silver (Ed.), *Teaching and Learning Mathematical Problem Solving: Multiple Research Perspectives* (pp. 361-379). Hillsdale, NJ: Erlbaum.

Secretary's Commission on Achieving Necessary Skills (1990). *SCANS Report.* Washington DC: U.S. Department of Labor.

Sfard, A., & Linchevski, L. (1994). "The gains and pitfalls of reflection: The case of algebra." *Educational Studies in Mathematics*, 26, 191-228.

Sfard, A. (1991). "On the dual nature of mathematical conceptions: Reflections on processes and objects as different sides of the same coin." *Educational Studies in Mathematics,* 22, 1-36.

Silver, E.A. (1994). "Dilemmas of mathematics instructional reform in the middle grades: The case of algebra." *QUASAR Occasional Paper.*

Steen, L.A. (1992). "Does everybody need to study algebra?" *Basic Education*, 37 (4), 9-13.

Swan, M. (1982). "The teaching of functions and graphs," in G. van Barneveld & H. Krabbendam (Eds.), *Proceedings of the Conference on Functions* (pp. 151-165). Enschede, The Netherlands: National Institute for Curriculum Development.

Thompson, P.W. (1995). "Quantitative reasoning, complexity, and additive structure." *Educational Studies in Mathematics.*

Thompson, P.W. (1994). "The development of the concept of speed and its relationship to concepts of rate," in G. Harel & J. Confrey (Eds.), *The Development of Multiplicative Reasoning in the Learning of Mathematics* (pp. 181-234). Albany: SUNY Press.

Thorpe, J.A. (1989). "Algebra: What should we teach and how should we teach it?," in S. Wagner & C. Kieran (Eds.), *Research Issues in the Learning and Teaching of Algebra* (pp. 11-24). Hillsdale, NJ: Erlbaum.

Tierney, C. & Monk, S. (in preparation). "Children's reasoning about change over time," in J. Kaput (Ed.), *Employing Children's Natural Powers to Build Algebraic Reasoning in the Content of Elementary Mathematics.*

U.S. Department of Labor (1987). *Work Force 2000: Work and Workers for the 21st Century.* "Executive Summary." Washington, DC: U.S. Government Printing Office.

University of Wisconsin-Madison & U.S. Department of Education (1993). In T.A. Romberg, E. Fennema, & T.P. Carpenter (Eds.), *Integrating Research on the Graphical Representation of Functions.* Hillsdale, NJ: Erlbaum.

University of Wisconsin-Madison & U.S. Department of Education (1991). In E. Fennema, T.P. Carpenter, & S.J. Lamon, *Integrating Research on Teaching and Learning Mathematics.* Albany: SUNY Press.

Usiskin, Z. (1988). "Conceptions of school algebra and uses of variables," in A.F. Coxford & A.P. Shulte (Eds.), *Ideas of Algebra, K-12* (pp. 8-19). Reston, VA: NCTM.

van Reeuwijk, M. (in preparation). "Algebra and realistic mathematics," in J. Kaput (Ed.), *Employing Children's Natural Powers to Build Algebraic Reasoning in the Content of Elementary Mathematics.*

Vergnaud, G. (1994). "Multiplicative conceptual field: What and why?," in G. Harel & J. Confrey (Eds.), *The Development of Multiplicative Reasoning in the Learning of Mathematics.* Albany: SUNY Press.

Wagner, S., & Kieran, C. (Eds.) (1989). *Research Issues in the Learning and Teaching of Algebra.* Reston, VA: NCTM.

Yerushalmy, M., & Schwartz, J.L. (1991). "Seizing the opportunity to make algebra mathematically and pedagogically interesting," in E. Fennema, T.P. Carpenter, & S.J. Lamon (Eds.), *Integrating Research on Teaching and Learning Mathematics* (pp. 41-68). Albany: SUNY Press.

Zawojewski, J.S. (1991). *Dealing with Data and Change Addenda Series, Grades 5-8.* Reston, VA: NCTM.